QH
499
R37
v.2

Regeneration in lower vertebrates
and invertebrates

P9-DXO-000

Regeneration In Lower Vertebrates And Invertebrates: II

Papers by
Margaret Egar, Charles Tweedle, Thomas L. Lentz, Roy A. Tassava, R. T. Francoeur, Sara Eisenberg-Zalik, J. C. Campbell, Randall W. Reyer et al.

MSS Information Corporation
655 Madison Avenue, New York, N. Y. 10021

Library of Congress Cataloging in Publication Data
Main entry under title:

Regeneration in lower vertebrates and invertebrates.

1. Regeneration (Biology). I. Eichler, Victor B.
QH499.R37 592'.03'1 72-8949
ISBN 0-8422-7052-3 (v. 3)

Copyright © 1972
by MSS Information Corporation
All Rights Reserved

QH
499
R37
v.2

TABLE OF CONTENTS

160727

CREDITS AND ACKNOWLEDGMENTS

Campbell, J. C.; and K. W. Jones, "The *in Vitro* Development of Lens from Cornea of Larval *Xenopus laevis*," *Developmental Biology*, 1968, 17:1-15.

Egar, Margaret; and Marcus Singer, "A Quantitative Electron Microscope Analysis of Peripheral Nerve in the Urodele Amphibian in Relation to Limb Regenerative Capacity," *Journal of Morphology*, 1971, 133:387-398.

Eisenberg-Zalik, Sara; and Violet Scott, "*In Vitro* Development of the Regenerating Lens," *Developmental Biology*, 1969, 19:368-379.

Francoeur, R. T., "General and Selective Inhibition of Amphibian Regeneration by Vinblastine and Dactinomycin," *Oncology*, 1968, 22:302-311.

Lentz, Thomas L., "Fine Structure of Sensory Ganglion Cells during Limb Regeneration of the Newt *Triturus*," *Journal of Comparative Neurology*, 1967, 131:301-322.

Reyer, Randall W., "DNA Synthesis and the Incroporation of Labeled Iris Cells into the Lens during Lens Regeneration in Adult Newts," *Developmental Biology*, 1971, 24:533-558.

Tassava, Roy A., "Hormonal and Nutritional Requirements for Limb Regeneration and Survival of Adult Newts," *Rhe Journal of Experimental Zoology*, 1969, 170:33-54.

Tassava, Roy A., "Survival and Limb Regeneration of Hypophysectomized Newts with Pituitary Xenografts from Larval Axolotls, *Ambystoma mexicanum*," *The Journal of Experimental Zoology*, 1969, 171:451-458.

Tweedle, Charles, "Transneuronal Effects of Amphibian Limb Regeneration," *The Journal of Experimental Zoology*, 1971, 177:13-30.

Zalik, Sara E.; and Vi Scott, "Development of ^{3}H-thymidine-labelled Iris in the Optic Chamber of Lentectomized Newts," *Experimental Cell Research*, 1971, 66:446-448.

PREFACE

Regeneration of limbs in amphibians and reptiles, regeneration of eye (Woffian regeneration) and regeneration of the basic body plan in Hydra are among the classic systems for the study of differentiation and growth.

Experimental investigation of these systems has proceeded at a high rate within the last five years, as papers in the present collection illustrate. These volumes, confined to the metazoa, consider the fundamental questions of morphogenesis including the possible storage and utilization of undifferentiated cells in the adult form, the means by which cells appear to "know where they are" in developing tissue, the process of dedifferentiation, and the control of regeneration by neural and endocrine factors. Intracellular regeneration, as in the ciliate protozoa, regeneration in plants, and regeneration in planaria including the Turbellarian flatworms are covered in separate volumes.

Neuronal Influence on Amphibian Limb Regeneration

A Quantitative Electron Microscope Analysis of Peripheral Nerve in the Urodele Amphibian in Relation to Limb Regenerative Capacity [1]

MARGARET EGAR AND MARCUS SINGER

Nerve fiber counts under the light microscope in silvered cross sections of the whole limb (Singer, '47a,b) and analysis of fiber sizes in the major limb nerves (Singer, Rzehak and Maier, '67) established the quantitative nature of the trophic control which peripheral nerves exert on limb regeneration in the newt, *Triturus (Notophthalmus) viridescens*.

Silver techniques are capricious; and there was no assurance that all the nerve fibers, particularly the fine unmyelinated ones, were counted. There is evidence in recent electron microscope studies that the number of fibers in the peripheral nerve may be much greater than traditional light microscope techniques have revealed because of the abundance, hitherto uncounted, of unmyelinated fibers [see Discussion]. The present work is a quantitative study of fibers in electron micrographs of the newt's peripheral nerve trunks. The results are compared to light microscope counts on the same nerves stained with silver for axons or with osmium for myelin sheaths. The results of the study, other than elucidating the quantitative morphology of peripheral nerves, allow a re-evaluation of the relation between number of nerve fibers and regenerative capacity of the limb.

MATERIALS AND METHODS

Spinal nerve 3 just distal to its ganglion, the brachial nerve in the upper arm, and the sciatic nerve in the upper leg were selected for study from three adult animals collected in western Massachusetts. Three millimeter sections of the nerves were removed from animals anesthetised with chloretone solution. Those from the left limbs were fixed in Bouin's, sectioned at 10 μ and stained according to Bodian's silver proteinate method to reveal the axons. The axons were projected and drawn with the aid of a Wild Zeichentube at a microscope magnification of 1000 times. The fibers were then counted by circling groups of ten.

[1] Supported by grants from American Cancer Society, The National Multiple Sclerosis Society and the National Institutes of Health.

Nerves from the right side of the same animals were exposed, fixed *in situ* with 1% osmic acid in phosphate buffer (Millonig, '61), removed in 1 mm lengths, and further fixed for two hours. A prolonged buffer wash was followed by rapid dehydration in graded ethanol and propylene oxide and then by araldite embedding.

Light microscope counts of myelinated nerve fibers and nuclei were made at × 900 magnification on 0.5 to 1.0 μ sections of the osmicated and araldite embedded tissue stained with toluidine blue (1% in borax). Photographs were compiled into montages for an additional check on the light microscope fiber count. Repeated counts indicated a less than 1% error.

Thin sections (500–600 Å) adjacent to the above thick sections were taken from the same blocks and mounted on one-hole formvar coated grids. Micrographs of individual nerves taken at initial magnifications of 1,600 to 10,000 were enlarged and combined into a montage of the entire cross-section from which all the fibers were counted.

RESULTS

Counts of myelinated fibers in osmicated sections under the light and electron microscope and in silvered sections with the light microscope are compared in table 1. There is an approximate one to one relation of osmicated myelinated fibers in the light and electron microscope sections. However, the number of axons counted in silvered sections exceeds the myelinated fiber values by about 23% on the average; the difference probably reflects bundles of unmyelinated fibers collectively stained with silver and not revealed by osmium.

The number of myelinated and unmyelinated fibers counted in electron microscope montages of the three nerves is recorded in table 2; the combined value for the two classes of fibers exceeds the number counted in silver sections (table 1). This suggests that the silver method distinguishes only a fraction of the unmyelinated fibers; perhaps unmyelinated fibers enwrapped by a single Schwann cell appear as one fiber under the light microscope. Therefore, a count was made of bundles of unmyelinated fibers in the electron microscope montages and this count was then combined with the myelinated fiber count (table 1). Such a combined value approximates the silvered fiber count more than it does the total fiber count, supporting the contention that silvered Bodian sections may stain the unmyelinated fibers as bundles.

The number of unmyelinated fibers approximates that of myelinated ones (table 2). Spinal nerve 3 favors the myelinated fibers (1 to 0.81–0.86); but sciatic and brachial nerves show a higher proportion of unmyelinated ones (1 to 1.06–1.51). The unmyelinated fiber count in the sciatic nerves is more variable than that in the other nerves (table 2). The distribution of the fascicles of unmyelinated fibers sharing a single Schwann investment was approximately random throughout the nerve cross-

TABLE 1

Number of fibers in three osmicated or silvered nerves

Nerve	Animal	Light microscope		Electron microscope	
		Osmium	Silver	Myelinated fibers plus unmyelinated bundles	Myelinated fibers
Sciatic	1	1127	1406	1480	1163
	2	1480	1165 [1]	1575	1277
	3	731	981	816	688
Brachial	1	816	959	1009	802
	2	869	1239	1032	813
	3	780	1006	1028	781
Spinal trunk (N. 3)	1	938	1112	1273	1053
	2	1104	1222	1341	1122
	3	885	1061	1050	870

[1] Count unreliable due to inadequate silver impregnation.

section. For this reason it would have been possible to estimate the total number of myelinated and unmyelinated fibers from counts of a sample part of the nerve. However, we elected to count the fibers in all of the nerve because of its small cross-section.

The cells seen in electron micrographs were predominantly Schwann cells (figs. 5, 6). These cells could be distinguished from fibroblasts (fig. 7) by their obvious association with the nerve fiber and by the fact that they possess a "basement" membrane, both criteria being absent in the fibroblast. Counts of fibroblasts in the montages yielded values ranging between one and nine whereas the Schwann cell number was 75 to 156 (table 3). Fibroblasts were most often superficially placed in the nerve, three being the largest number of fibroblasts located centrally in any fascicle studied. Counts were also made of Schwann cell nuclei for the two fiber classes. The results were an average of 93 for myelinated and seven for unmyelinated fibers. Thus the number associated with unmyelinated axons (fig. 6) constitutes less than 10% of the total.

The present study was stimulated by interest in the neurotrophic control which nerves exert on limb regeneration. Previous studies indicated that neurotrophic activity

TABLE 2

Numbers and ratios of unmyelinated (U) and myelinated (M) fibers in electron micrographs of three nerves

Nerve	Animal	Myelinated fibers	Unmyelinated fibers	Total	Ratio of myelinated to unmyelinated
	1	1163	1281	2444	1:1.1
Sciatic	2	1277	1931	3208	1:1.51
	3	688	833	1521	1:1.5
	1	802	857	1659	1:1.06
Brachial	2	813	868	1681	1:1.06
	3	781	894	1675	1:1.14
Spinal trunk (N. 3)	1	1053	903	1956	1:0.86
	2	1122	912	2034	1:0.81
	3	870	708	1578	1:0.81
				Average ratio	1:1.095

TABLE 3

Number of nuclei in Triturus nerves

	Animal	Light microscope	Electron microscope			
		Total nuclei	Total nuclei	Nuclei of myelinated fibers	Nuclei of unmyelinated fibers	Nuclei of fibroblasts
	1	153	131	120	7	4
Sciatic	2	166	155	147	9	9
	3	84	80	74	3	3
Average		134	122	114	6	5
	1	97	82	72	9	1
Brachial	2	89	78	69	6	3
	3	86	76	68	7	1
Average		92	79	70	7	2
Spinal trunk (N. 3)	1	142	119	103	11	5
	2	103	100	91	8	1
	3	97	98	92	3	3
Average		114	106	95	7	3
Average per cent for all nerves				90%	7%	3%

Note: Counts made on the total cross-section of nerve.

is directly related to the amount of neuroplasm available at the amputation surface of the limb (Singer, Rzehak and Maier, '67). The amount of neuroplasm is a function of the number of axons and the average cross-sectional area of the axons. The earlier studies on the quantitative relations between the nerve and its influence on limb regeneration were based on the silver stain. Since the results presented here show that the total number of unmyelinated axons cannot be counted in silver preparations, a more accurate estimate of the amount of neuroplasm available for regeneration activity is found in the electron micrographs (fig. 4). An average value for the diameter of unmyelinated and myelinated axons was obtained by measuring at random 20 myelinated and 20 unmyelinated fibers of the sciatic nerve. When an axon was irregular in outline, the shortest and longest diameters were averaged. The cross-sectional area was then calculated from the average value for the two classes of fibers. The results of measurements and calculations are recorded in table 4. The neuroplasmic contribution of the unmyelinated fiber is small, less than 3% of the total neuroplasm of the nerve, although the number of these fibers is approximately equal to that of the myelinated ones.

TABLE 4

Diameters of 20 myelinated and 20 unmyelinated fibers of the sciatic nerve [1]

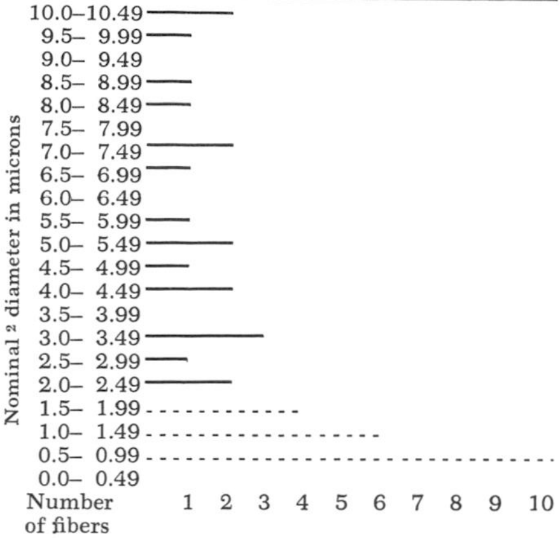

Nominal [2] diameter in microns	Number of fibers: Myelinated fibers (———)	Number of fibers: Unmyelinated fibers (- - - -)
10.0–10.49	2	
9.5– 9.99	1	
9.0– 9.49		
8.5– 8.99	1	
8.0– 8.49	1	
7.5– 7.99		
7.0– 7.49	2	
6.5– 6.99	1	
6.0– 6.49		
5.5– 5.99	1	
5.0– 5.49	2	
4.5– 4.99	1	
4.0– 4.49	2	
3.5– 3.99		
3.0– 3.49	3	
2.5– 2.99	1	
2.0– 2.49	2	
1.5– 1.99		4
1.0– 1.49		6
0.5– 0.99		10
0.0– 0.49		

[1] Measurements made on an electron micrograph montage (× 5,000).

[2] Nominal = (greatest + least)/2.

Myelinated fibers ——— (average area = 31.48 microns ^{2})

Unmyelinated fibers - - - - - - (average area = 0.89 microns ^{2})

DISCUSSION

Electron micrographs offer a new dimension in quantitative analyses of the nervous system. Earlier light microscope analyses of peripheral nerve fibers were made on silvered or osmicated preparations. These two preparations for counting axons yielded considerable differences which were in part attributed to unmyelinated fibers and in part to the capriciousness of silver techniques. Hughes ('69) in his review of quantitative aspects of neurohistology has summarized these findings on cranial as well as peripheral nerves.

The difficulties in counting unmyelinated axons, particularly the smaller ones, are illustrated in the article by Gasser ('55) which describes electron micrographs of nerve stained en bloc with silver and then embedded in plastic for electron microscopic study. Silver was deposited diffusely around a bundle of unmyelinated fibers without revealing individual axons. Maturana's ('59) ultrastructural study of the frog optic nerve indicates that earlier silver and osmium counts underestimate the unmyelinated fiber number by a factor of 30 or more. Our experiences conform with the conclusion that a bundle of unmyelinated fibers is impregnated as a unit. Sampling techniques employed in the counting procedures were also responsible for the great variation in the ratios.

Electron microscopy has eliminated many uncertainties in fiber counts and has provided a means for determining accurately the number of each type of fiber with the one salient variable of the sampling technique. The electron microscopic method of quantitative analysis is tedious and exacting particularly when montages are employed for total fiber count at sufficiently high magnification. The use of montages has been reported by few workers: in crustaceans by McAlear, Camougis and Thibodeau ('61), Nunnemacher ('66), Sutherland and Nunnemacher ('68), Skobe and Nunnemacher ('70), Somers and Nunnemacher ('70); in *Xenopus* by Gaze and Peters ('61); in the teleost by Westerman and Wilson ('68); and in the mouse and roach by McAlear, Camougis and Thibodeau ('61). Friede and Samorajski ('67) estimated the fiber value in mice by calculating the total number from sampled areas of a cross-section. They reported that the ratio of myelinated to unmyelinated fibers is about one to three in the sciatic nerve and about one to four and one-half in the vagus nerve. In our counts on montages of *Triturus* peripheral nerve the ratio is approximately one myelinated to one unmyelinated fiber. Moreover, our studies show that Bodian silvered preparations represent the myelinated fiber count more than the total fiber number. Only a small fraction of unmyelinated fibers are revealed by this method; and as noted this fraction reflects the number of bundles of unmyelinated fibers rather than the total number of fibers.

The fiber count in electron micrographs was of particular interest in light of our previous studies on the relation between number of fibers and neurotrophic control of limb regeneration. In previous reports from this laboratory (see review Singer, '52) the total number of nerve fibers at the amputation surface of the upper arm was reported to be approximately 2500 in silvered sections. If we assume that the present results on the relative number of myelinated to unmyelinated fibers in the major nerve trunks also obtains for the smaller branches and the scattered nerve fibers, then the total fiber number at the amputation surface should be approximately doubled to obtain a truer estimate of the innervation. It may be that the real number is even greater since more unmyelinated fibers might be expected among the smaller nerve branches and the scattered fibers approaching their terminations.

This new estimate of the quantity of limb innervation makes it now possible to calculate more accurately the threshold number of nerve fibers required to initiate limb regeneration. Previously we recorded a threshold value of about 1040 nerve fibers to induce regrowth of the upper limb, or expressed per unit amputation area, 9.6 fibers per 100 μ^2 (Singer, '47b). Assuming that the one to one ratio of myelinated to unmyelinated fibers established in the present study obtains in fact, the threshold number should be about 2000 fibers or about 20 fibers per 100 μ^2. Counts made on more distal parts of the forelimb (see table 2 of Singer, '47a) should be similarly corrected. These reassessments do not alter the major concept drawn from the previous quantitative studies, namely that a threshold quantity of nerve fibers is required to satisfy the neurotrophic needs of limb tissues to regenerate.

Other studies from our laboratory indicated that the number of nerve fibers is a lesser measure of the neurotrophic contribution to the regeneration process than is the amount of neuroplasm represented by these nerve fibers (Singer, Rzehak and Maier, '67). In the present study we calculated the average amount of neuroplasm available in cross-sections of unmyelinated and myelinated fibers. The results showed that unmyelinated fibers by virtue of their small size contribute only about 3% of the neuroplasmic area of the nerve cross-section. If our previous conclusion is correct, namely that the neurotrophic contribution of the axon is directly related to its cross-sectional size (volume), then the unmyelinated fibers contribute only a few per cent of the neurotrophic effect. Since the number of myelinated fibers required to satisfy threshold needs for limb regeneration is approximately 1000, by previous silver counts, a trophically equivalent number of unmyelinated fibers should be about 33 times the number, or 33,000 fibers. Stated otherwise, the average myelinated fiber in the newt is about 33 times as effective an agent of limb regeneration as the average unmyelinated fiber.

It must be emphasized that the neurotrophic effectiveness of a fiber is related not to the presence or absence of myelin but rather to the size of that fiber (Singer et al., '67). Thus, unmyelinated fibers of large cross-sectional area should be more effective agents of growth than small unmyelinated fibers.

Finally, we wish to comment briefly upon the counts made of Schwann cell bodies and of fibroblasts in the electron micrograph montages. Previous workers who have counted fibroblasts in light microscope sections of nerve have presented values ranging as high as 39% (Abercrombie and Johnson, '46; Thomas, '48). Causey

('60) discusses the earlier findings and concludes from electron microscope work that fibroblasts contribute less than 5% of the cells in nerves. Our findings of 3% fibroblast nuclei inside the perineurium support his conclusion. Moreover, most of the fibroblasts were peripherally located in the bundles. No fascicle had more than three fibroblast nuclei deeper than the usual position adjacent to the inner surface of the perineurium.

ACKNOWLEDGMENT

The authors wish to express their appreciation to Mrs. Merry A. Harrison for her assistance.

LITERATURE CITED

Abercrombie, M., and M. L. Johnson 1946 Quantitative histology of Wallerian degeneration. I. Nuclear population in rabbit sciatic nerve. J. Anat., *80:* 37–50.

Causey, G. 1960 The Cell of Schwann. E. and S. Livingstone Ltd. Edinburgh and London, 120 pp.

Friede, R. L., and T. Samorajski 1967 Relation between the number of lamellae and axon circumference in fibers of vagus and sciatic nerves of mice. J. Comp. Neur., *130:* 223–232.

Gasser, H. S. 1955 Properties of dorsal root unmedullated fibers on the two sides of the ganglion. J. Gen. Physiol., *38:* 709–728.

Gaze, R. M., and A. Peters 1961 The development structure and composition of the optic nerve of *Xenopus laevis* (Daudin). Quart. J. Exp. Physiol., *46:* 299–309.

Hughes, A. F. W. 1969 Quantitative aspects of neurohistology. Int'l. Rev. Gen. Exp. Zool., *4:* 169–217.

Maturana, H. R. 1959 Number of fibers in the optic nerve and the number of ganglion cells in the retina of Anurans. Nature, *183:* 1406.

McAlear, J. H., G. Camougis and L. F. Thibodeau 1961 Mapping of large areas with the electron microscope. J. Biophys. Biochem. Cytol., *10:* 133–135.

Millonig, G. 1961 Advantages of a phosphate buffer for OsO_4 solutions in fixation. J. Applied Physics, *32:* 1637.

Nunnemacher, R. F. 1966 The fine structure of optic tracts of Decapoda. In: International Symposium of the Functional Organization of the Compound Eye. C. G. Bernhard, ed. Pergamon Press Ltd. London, *7:* 363–375.

Singer, M. 1947a The nervous system and regeneration of the forelimb of adult *Triturus* VI. A further study of the importance of nerve number, including quantitative measurements of limb innervation. J. Morph., *104:* 223–249.

——— 1947b The nervous system and regeneration of the forelimb of adult *Triturus* VII. The relation between number of nerve fibers and surface area of amputation. J. Morph., *104:* 251–265.

——— 1952 The influence of the nerve in regeneration of the amphibian extremity. Quart. Rev. Biol., *27:* 169–200.

Singer, M., K. Rzehak and C. S. Maier 1967 The relation between the caliber of the axon and the trophic activity of nerves in limb regeneration. J. Exp. Zool., *166:* 89–98.

Skobe, Z., and R. F. Nunnemacher 1970 The fine structure of the circumesophageal nerve in several decapod Crustaceans. J. Comp. Neur., *139:* 81–91.

Somers, M. E., and R. F. Nunnemacher 1970 Microanatomy of the ganglionic roots of the abdominal cord of the crayfish, *Orconectes virilis* (Hagen). J. Comp. Neur., *138:* 209–218.

Sutherland, R. M., and R. F. Nunnemacher 1968 The microanatomy of crayfish thoracic cord and roots. J. Comp. Neur., *132:* 499–518.

Thomas, G. A. 1948 Quantitative histology of Wallerian degeneration. II. Nuclear populations in two nerves of different fibre spectrum. J. Anat. London, *82:* 135–145.

Westerman, R. A., and J. A. F. Wilson 1968 The fine structure of the olfactory tract in the teleost *Carassius carassius* L. Z. Zellforsch., *91:* 186–199.

EXPLANATION OF FIGURE

1 An electron micrograph montage of a spinal nerve 3 reduced from an original size of 3 by 2 m. The reduction gives a false notion of the clearness with which unmyelinated fibers may be seen. Magnification × 750.

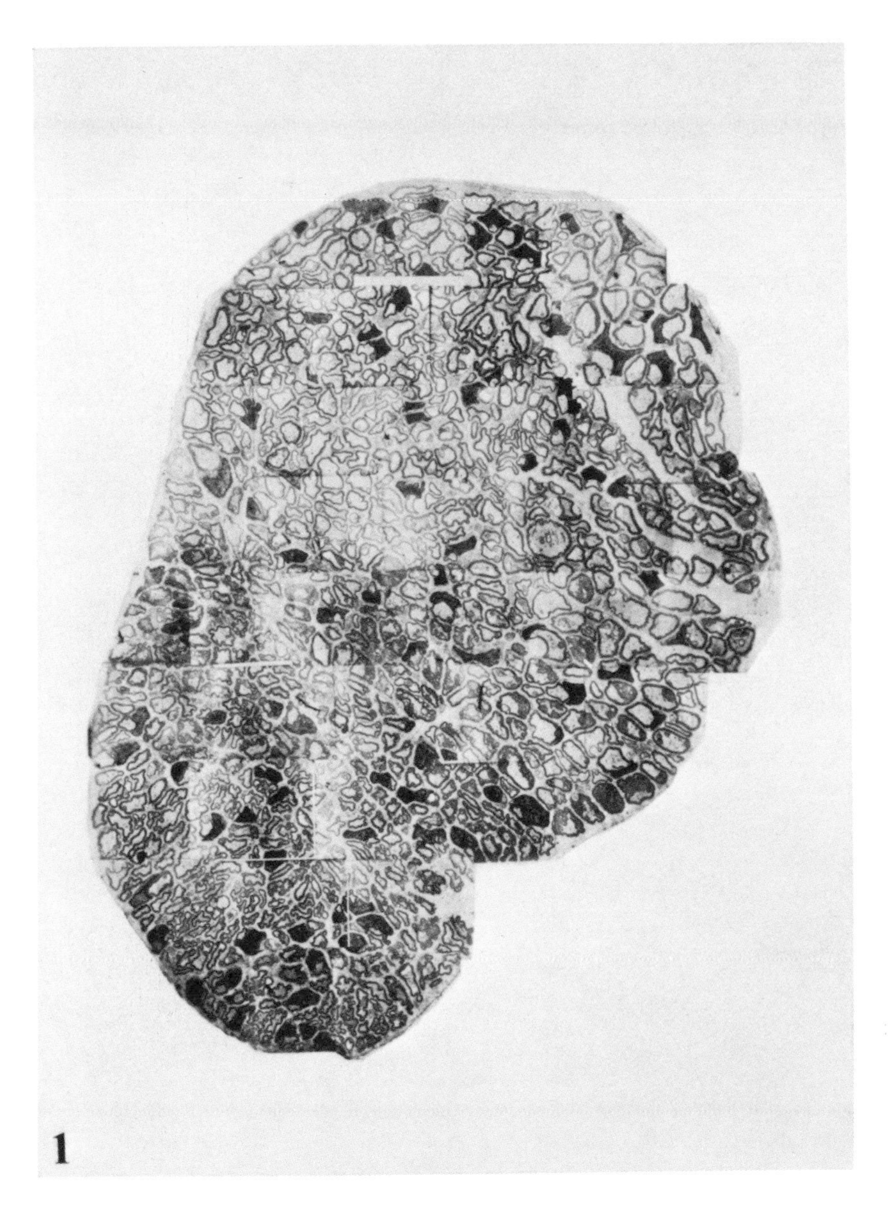

PLATE 2

EXPLANATION OF FIGURES

2 The left brachial nerve stained for silver. Magnification × 1200.

3 A thick section (1.0 μ) of the osmicated right brachial nerve used in light microscope counts of myelinated fibers. Magnification × 1200.

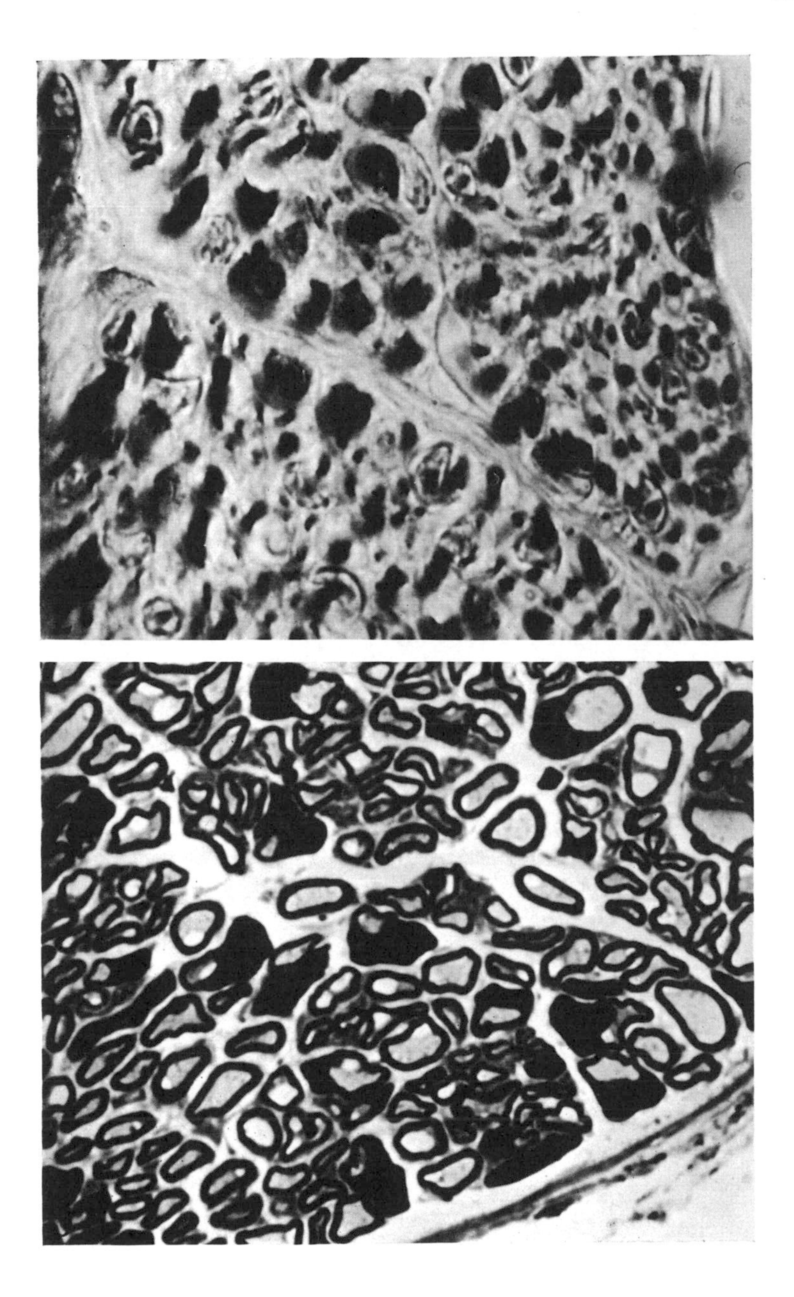

PLATE 3

EXPLANATION OF FIGURES

4 A reproduction at a higher magnification of part of a brachial nerve montage to illustrate the nature of the micrographs from which counts were made. Myelinated axon (A); myelin sheath (M); unmyelinated axon (a); "basement" membrane (arrow, b); endoneurial collagen fibers (ec). Magnification × 20,000.

5 A portion of spinal nerve 3 showing the Schwann nucleus (SN) of a myelinated fiber. The arrow (b) indicates the "basement" membrane that is characteristic of Schwann cells but not of the fibroblast (F). Axon of a myelinated fiber (A); endoneurium (e). Magnification × 10,500.

6 A Schwann cell surrounding unmyelinated axons from spinal nerve 3. Notice that such a cell also has a "basement" membrane (arrow, b). Unmyelinated axon (a); myelin sheath (M); Schwann nucleus (SN). Magnification × 20,000.

7 A fibroblast from deep within a spinal nerve 3 fascicle. Notice the long cytoplasmic projection (arrow) and the absence of a "basement" membrane. Myelin sheath (M). Magnification × 6,400.

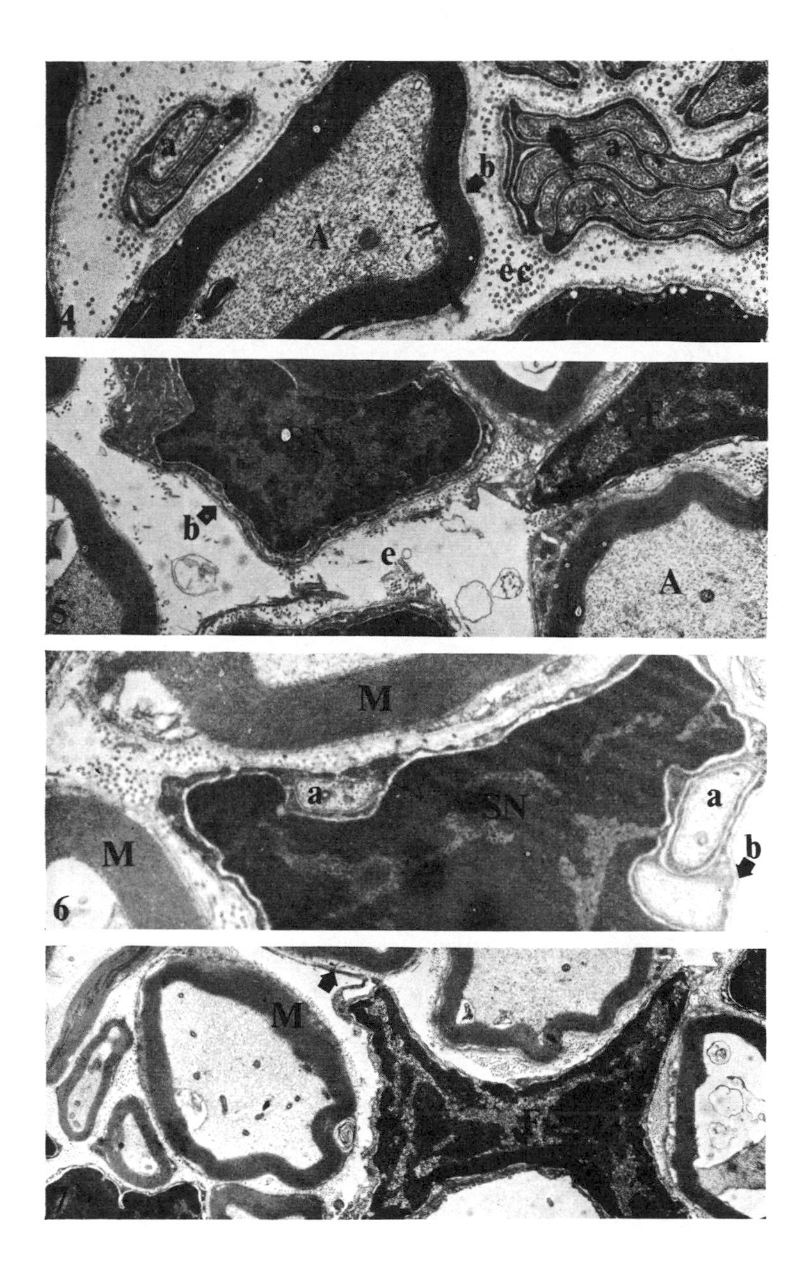
a
a
b
A
ec
4
b
e
A
5
M
a
SN
a
M
b
6
M
7

Transneuronal Effects on Amphibian Limb Regeneration

CHARLES TWEEDLE

Students of regeneration often make the tacit assumption that a regenerating limb in an animal is not influenced by other regenerating limbs in the same animal. Experiments have commonly been designed so that one limb of the animal is used as a control while another limb of the same animal receives some experimental treatment. This practice makes no allowance for any factor of interaction that might exist with the removal of more than one limb.

T. H. Morgan ('06) reported from unaided visual observations that the removal of one, two, or three limbs from adult *Diemyctylus viridescens* did not seem to alter the rate of regeneration. A more elaborate experiment focused on this problem was carried out by Zeleny in 1909. As part of this study he used *Ambystoma opacum* larvae with different numbers of limbs removed and sometimes also the tail. Although he concluded that the larvae with the greater "degree of injury" (more appendages removed) had an increased regeneration rate, modern statistical analysis of his data does not bear out his conclusions. Not only is there great overlap between regeneration rates of the various groups but the small number of animals used precludes statistical significance.

There have also been a number of studies done on the effect of hindlimb removal on tail regeneration. Zeleny ('09) studied the effect of the removal of one or both hindlimbs on the rate of tail regeneration in *R. clamitans* tadpoles and found that, while the amputation of one hindlimb had the effect of accelerating tail regeneration, the removal of both hindlimbs brought about a weaker acceleration of the regeneration rate. Zeleny concluded from this that there was an optimal amount of injury to stimulate regeneration; more than this optimum brought about a lessened stimulation. However, in 1926 Hiller repeated these experiments with axolotls and reported slower tail regeneration with the simultaneous removal of limbs as well as the tail. In an attempt to settle this problem, Blacher et al. ('32) carried out extensive studies with tadpoles on the effect of limb amputation on the tail regenerate in order to determine whether there was any time relationship with the effects of the additional amputation. They found that the removal of the hindlimbs stimulated tail regeneration, but only if the tail was amputated not more than two days before the limbs were removed.

There are a number of reports of regeneration promoters that may be active in the growth stages of regeneration. Issekutz-Wolsky and Fogarty ('62) demonstrated that a tail blastema implanted beneath the skin of the lateral body surface stimulated tail regeneration in newts. The authors also reported that neither implantation of amputated normal tail ends nor infliction of wounds brought about accelerated regeneration. Skowron et al. ('63) found acceleration of regenerative growth in axolotls with not only the implantation of whole regeneration blastemata beneath the skin but also with the implantation of muscles, injections of dissociated blastemal cells or of homogenates from such cells. Weber and Maron ('65) have also reported increased regeneration after the implantation of stump muscle under the abdominal skin in axolotls where it dedifferentiated into blastema-like cells. They localized the promoter activity mainly in the microsomal fraction which contained soluble proteins; when this was injected into the body wall of axolotls with amputated tails, an accelerated regeneration rate was noted.

Work by Twitty and Delanney ('39) has shown that competition for circulating nutrients did not affect the speed of size regulation of eyes transplanted from younger to older *Ambystoma* larvae with regenerating tails. From these and other experiments the authors concluded that there is a moderate abundance of nutritive materials in circulation even when the animals are starved. These are gradually depleted but apparently not to the point where competition becomes a vital factor in their distribution. Thus, if one may extrapolate somewhat, it seems unlikely that this should be an important consideration in a well-fed animal with more than one regenerating limb.

The growth and maintenance of each organ depends to a considerable extent on its innervation (Wyburn-Mason, '50). If a limb is denervated its growth is seriously retarded, particularly if reinnervation is prevented. On the other hand, an augmentation of the normal nerve supply to a limb may cause great hypertrophy or the production of supernumerary limbs (Huxley and De Beer, '34). The trophic effect of the peripheral nerve is also seen in regenerating limbs (Singer, '52, '59, '60) so much that the limbs of adult Anura which do not regenerate normally will do so with extra nerves added. If regenerating limbs are denervated, their growth ceases until they become reinnervated. Thus, it is seen that any change in the nerve supply can also be a factor in the rate of regeneration.

It was not clear which of these factors, if any, might be of importance in altering the rate of limb regeneration when more than one limb in an animal is removed. Therefore the experiments described herein were designed to determine the effect of amputation of an extra limb and then to assess the cause of any effect found.

It was found that removal of a second limb brought about a slower regeneration rate, but only if the two limbs removed were directly opposite one another. Evidence is presented that a contralateral neural injury is responsible for this effect.

MATERIALS, METHODS, AND RESULTS

General methods

Larval axolotls, *Ambystoma mexicanum*, used in these experiments were obtained through the generosity of Dr. R. R. Humphreys at Indiana University. The *Ambystoma maculatum* used in the aneurogenic experiment were obtained from Glen Gentry, Donelson, Tennessee. Adult newts, *Notophthalmus (Triturus) viridescens*, were obtained from Lewis Babbitt, Petersham, Massachusetts. All experimental animals were fed every other day on sliced beef liver. The animals were maintained at a uniform laboratory temperature of 22 ± 2°C. All amputations were made through the mid-humerus following anesthesia in 1:2000 MS 222 (Sandoz). In all amputations the bone was carefully trimmed with iridectomy scissors to avoid interference with regeneration. Care was taken to make certain that the amputations were as nearly as possible at the same level on the limb, since level of amputation has been reported to have an effect on the rate of regeneration in a number of regenerating systems. In Amphibia in general the absolute rate of regeneration increases with the percentage of limb removed (Needham, '52). To check this factor the length of limb stump remaining after amputation was measured in each case and plotted against amount of linear regeneration at various time intervals. No correlation was found with the amputation levels used.

Length measurements were used as an index of regenerative growth. These measurements were made with a dissecting microscope fitted with an optical micrometer. Following anesthesia the animals were positioned on their backs on moist gauze on the microscope stage so as to achieve accurate and reproducible linear measurements. Repetition of these measurements showed the results to be consistent. Each measurement was made at least twice and the average used. Amount of linear regeneration was determined as the amount of growth in length beyond the length of the amputation stump. For measurements of linear growth the total length of the regenerating limb was taken and the length of the amputation stump subtracted to give the amount of new growth. Observations were also made on morphogenesis to supplement the length measurements.

SERIES 1

Growth relationships between single and double limb amputations

Procedure

Fifty-five laboratory-raised *Ambystoma mexicanum* larvae of 60 mm in length were used. These were maintained in individual small plastic dishes. The animals were divided into five groups on the basis of amputation procedure. These groups were as follows:

A. one forelimb removed; B. one hindlimb and one forelimb removed; C. one hindlimb removed; D. both hindlimbs removed; E. both forelimbs removed.

In order to satisfy the independence requirements in statistics, in the groups with two hindlimbs or two forelimbs removed only one of the limbs was arbitrarily chosen for measurement. In the groups with one hindlimb and one forelimb removed only the forelimb regenerate was measured for the same reason. Length determinations were made only until shortly after the last of the experimental animals had reformed all digits (about 45 days post-amputation). After this time, the growth of the digits made both measurements and interpretation difficult.

Results

Figure 1 shows a plot of the mean regenerate length of each of the five groups plotted against time after limb amputation. If the graph length is given in micrometer units, each micrometer unit is equal to 0.15 mm. About day 20 there develops a difference among the experimental classes whereby the animals with both hindlimbs or both forelimbs amputated show a slower linear regeneration rate than the other three groups. After ascertaining compliance with assumptions of the statistical tests, a one-way analysis of variance and new multiple range test were run at 35 days post-amputation and found to show, indeed, that the two single limb amputations and the single hindlimb plus single forelimb groups display statistically greater ($P < 0.01$) growth by 35 days after amputation than the other two groups. The morphogenetic stages closely paralleled the growth results. Table 1 presents the time of complete digit formation in the five groups.

Just as the animals with single forelimb or single hindlimb amputation undergo greater regeneration by 35 days than those with double forelimb and double hindlimb amputations, it is also seen that the latter two groups show a delayed onset of morphogenesis. The forelimb of the group with both a hindlimb and forelimb removed, however, showed a morphogenetic rate similar to the single forelimb amputation. It is interesting that the hindlimb, which undergoes as rapid growth in length as the forelimb, lags behind it in morphogenesis. This is probably due to the greater size of the hindlimb. Thus, both linear growth and morphogenesis were found to be delayed with the removal of a second limb *but only if the second limb removed were directly contralateral to the first limb amputated.* This would seem to deny a systemic or

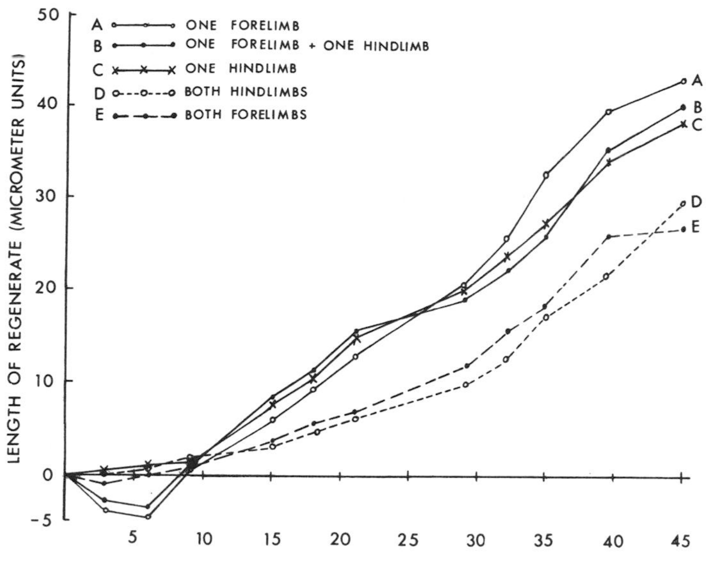

Fig. 1 A comparison of the mean linear regeneration length of limbs of *Ambystoma mexicanum* larvae following 1 of 5 amputation procedures: A. one forelimb removed; B. one hindlimb and one forelimb removed; C. one hindlimb removed; D. both hindlimbs removed; E. both forelimbs removed. Variance analysis and a new multiple range test run at 35 days after amputation showed that the animals in groups A,B, and C displayed significantly ($P < 0.01$) greater limb growth than the other two groups.

nutritional effect, otherwise the removal of one hindlimb plus one forelimb would have also brought about a slower regeneration. In summary, there exists a localized effect on regenerative growth whereby the removal of two contralateral limbs brings about a slower linear regeneration rate than single limb amputation.

SERIES 2

Cytochemical investigation of transneuronal effects in ambystoma

The results in Series 1 imply a local inhibition of regenerative growth with removal of two contralaterally-opposed limbs. These results were reminiscent of neuroembryological studies by Detwiler ('36) on *Ambystoma* embryos which showed that the removal of both forelimb buds brought about a greater reduction in the volume of the developing motor roots of the brachial nerves than when only one forelimb was removed. This was thought to be due to a large complement of commissural neurons which not only made dendritic contact throughout the white substance on the same side of the spinal cord but also cross to the opposite side via the ventral commissure. Such neurons have been described in salamanders by Herrick and Coghill ('15) and earlier by van Gehuchten (1898). The present series was designed to determine whether transneuronal effects might also be involved in the inhibition of growth in larval *Ambystoma* regenerates.

Transneuronal degenerative effects will here be used to mean those which are produced on a nerve cell by being in contact with another nerve cell which has been injured. These transneuronal effects have the same characteristics as the chromatolytic reaction of neurons to direct injury (Cole, '68). There is a large literature concerning chromatolysis of a nerve cell following axonal transection (see Ducker et al., '69, for review).

TABLE 1

A comparison of the time of complete digit formation in Ambystoma mexicanum larvae following various amputation procedures

	Amputation group	Days after amputation				
		33	36	39	42	45
% Complete digit formation	One forelimb N = 9	90	100	100	100	100
	Both forelimbs N = 8	38	50	50	62	100
	One hindlimb N = 10	40	70	100	100	100
	Both hindlimbs N = 18	5	16	40	62	100
	Forelimb of hindlimb + forelimb group N = 10	70	100	100	100	100

Procedure

Small blocks of tissue from the brachial spinal cord region of *Ambystoma mexicanum* larvae and adult *Notophthalmus viridescens* were fixed either in Zenker's or Carnoy's at intervals following removal of one, both or no forelimbs. Embedding was both in glycol methacrylate by the method of Ashley and Feder (unpublished) and in paraplast. Sectioning was done in both cases on a regular mechanical microtome. Two micron thick methacrylate sections were stained with methylene blue (1%) or Azure B (0.25% at pH 4.2). After dehydration the paraplast sections were stained with gallocyanin (Berube et al. '66) or Azure B.

Feulgen staining was also done on both methacrylate sections and 7 μ paraplast sections according to the technique of Kasten and Burton ('59). Staining was also carried out for degenerating nerve fiber endings by the method of Fink and Heimer ('67).

Results

Although characteristic changes in Nissl substance could be seen following axonal transection, it was discovered that another useful morphological criterion of the neuronal disruption is that of changes in the chromatin of the nuclei. In the chromatolytic nerve cells the chromatin was seen to be more aggregated than in control neurons (see figs. 2a,b). This was especially noticeable in the Feulgen preparations. Similar findings were reported by Lentz ('67) in an electron microscopic study of the response of the sensory ganglia of the newt to limb amputation. He reported that within the nucleus the granular and fibrillar material becomes more aggregated, occurring in clumps. The great reproducibility of these nuclear changes following different fixations (Carnoy's, formalin, Bouin's, and Zenker's) precludes the possibility of a fixation artifact. After single limb amputation, these nuclear changes were seen in neurons on both sides of the spinal cord, although less on the side opposite the transection. This occurred both in *Ambystoma* and *Triturus*. In the 2 μ thick methacrylate sections and in *Ambystoma* in particular chromatin clumping was also seen in the ganglia on *both* sides of the animal following unilateral limb removal, although the degree of nuclear change was much less in the ganglia contralateral to the limb amputation. The neurons most obviously affected by unilateral limb removal were the cells of the dorsal and ventral portions of the spinal cord and ganglia of the side of limb removal; however, there were detectable effects on just about every neuron in the spinal cord. It was difficult to ascribe all the changes seen to known neural pathways along which transsynaptic degeneration might pass. There were neurons showing definite nuclear changes on the side of the animal opposite the amputation that would have no direct synaptic connections to the operated side — at least according to the scheme laid down by van Gehuchten (1898). This same phenomenon has been reported by others such as Ochs et al. ('60) who found chromatolysis in lateral nuclear neurons contralateral to the side of unilateral sciatic nerve crush in the cat. Here, also, no known neural connections could explain the results. In *Ambystoma* these nuclear changes were seen in the spinal cord and

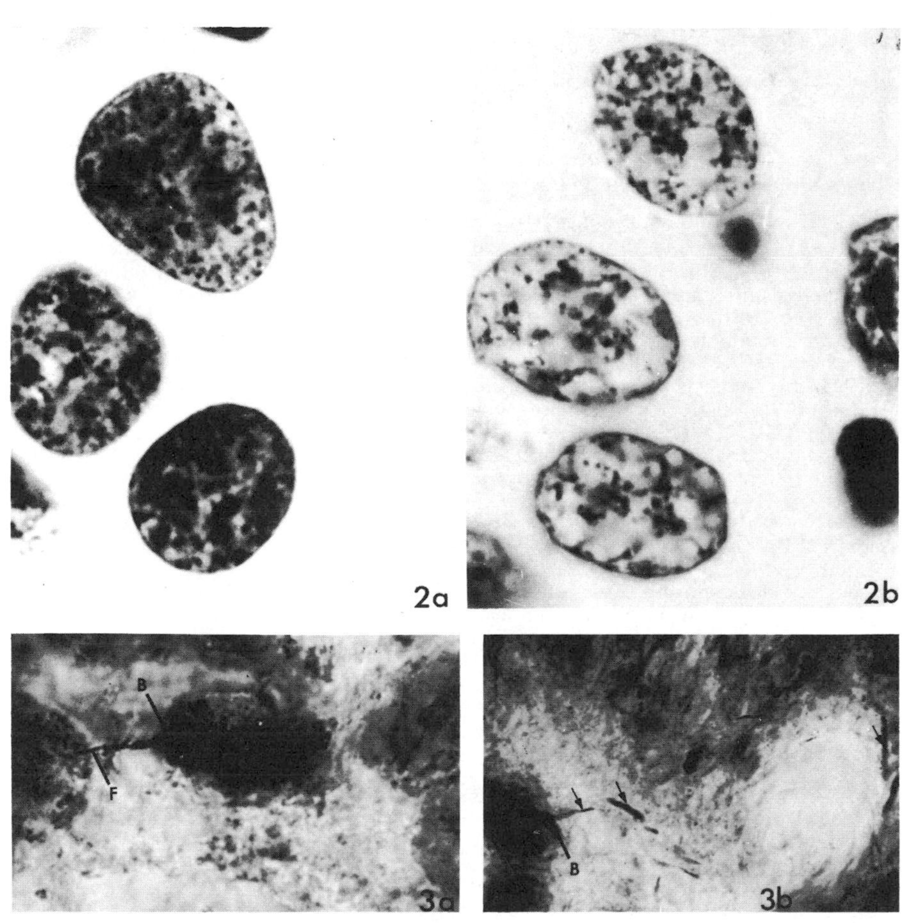

Fig. 2a Light micrograph of a Feulgen preparation of normal ventral spinal cord nuclei in an unamputated animal. Note the relatively diffuse state of the chromatin. (× 3500)

Fig. 2b Light micrograph of a similar Feulgen preparation of ventral spinal cord nuclei at nine days after limb amputation. The chromatin is seen to change from the more diffuse appearance of the normal nuclei to a more aggregated appearance in the chromatolytic neuron. The nuclear changes were seen on both sides of the spinal cord after single limb removal. (× 3500)

Fig. 3a Light micrograph of a cross section of a ventral spinal cord cell showing a degenerating nerve fiber ending (F) and bouton (B). Degenerating fibers were seen on both sides of the spinal cord following single limb amputation, giving evidence for transneuronal degeneration. (× 1500)

Fig. 3b Light micrograph of a cross section of the ventral spinal cord region at 15 days after single forelimb amputation, showing a number of degenerating nerve fiber endings (arrows) and a degenerating bouton (B). (× 1250)

ganglion cells by one day after nerve section and lasted about 10-12 days in the case of single forelimb amputation and 15-18 days in the case of the double forelimb amputation. It is thought that the lengthier disruption seen in the animals with double forelimb amputation is due to the fact that not only must the nerve cells recover from the direct injury of having their axons severed with limb removal but also from contralateral changes carried from the injured neurons of the opposite side of the spinal cord. The phenomenon of chromatin aggregation is especially interesting in the light of a report by Causey and Strattman ('59) who presented microspectrophotometric evidence that there is a loss of Feulgen staining material (DNA) during chromatolysis in the cell bodies of the cervical ganglion of the rabbit. Preliminary studies (Tweedle, unpublished) show a similar "loss" of Feulgen positive material in the case of chromatolytic brachial ganglia of *Ambystoma*. However, it is debatable whether this represents a loss in DNA. Noeske ('69) has recently presented evidence that, as a consequence of chemical and biophysical alterations in chromatin during different functional states, variations in Feulgen values may be found in cell nuclei that really contain the same amount of DNA.

Another index of neural injury is the presence of degenerating nerve fiber endings or synapses (see Gray and Guillery, '66, for review). The technique of Fink and Heimer ('67) was used to stain for degenerating axonal endings in *Ambystoma* at 7 and 15 days after unilateral limb removal in order to detect the presence of neural pathways that might be involved in the contralateral changes seen. Sections from unamputated animals were also run through the stain as a control. It was found that unilateral limb removal brought about degenerating fiber endings in both the dorsal and ventral areas of the spinal cord and occasionally in the ventral area of the spinal cord and occasionally in the ventral area of the unamputated side (see figs. 3a,b). The degenerating fiber endings were mostly noticeable in the spinal cord at the level of the brachial ganglia, which supports the idea that they are indicative of reflex pathways between the forelimbs. The appearance of degenerating axonal endings in the ventral portion of the unamputated side of the spinal cord would argue for the involvement of transneuronal injury across one extra neuron — the commissural fibers from the dorsal area of the amputated side to the ventral area of the contralateral side. This type of transneuronal disruption is not rare in young animals and in neural pathways with few collaterals (Bleier, '69).

It is interesting that although degenerating fibers are seen, there is no great evidence of nerve cell death following limb amputation. This may be related to the fact that there is known to be some degree of regeneration in the amphibian CNS (Clemente, '64) or the fact that long-term observations on the brachial neurons were not made following limb amputation. In summary, the results of Series 2 demonstrate that following unilateral limb amputation there are bilateral changes such that brachial neurons on both sides of the spinal cord are affected. With double limb removal the chromatolytic appearance of the brachial neurons (both in the spinal cord and the ganglia) is prolonged for a greater period time than with single limb removal; this is apparently due in part to transneuronal degeneration.

SERIES 3

Autoradiographic analysis of contralateral effects following limb amputation

The results of Series 2 gave morphological evidence for transneuronal changes following limb amputation. Series 3 is designed to give a more quantitative evaluation of this disruption. Advantage was taken of the fact that chromatolytic neurons have been found to increase RNA production during the earlier stages of the nerve reaction (Haddad et al., '69; Watson, '65; Lison, '62).

Procedure

Uridine-5-^{3}H (s.a. greater than 20 c/mM, New England Nuclear) was used as a precursor of RNA and the incorporation of the label into the spinal cord and brachial ganglia of *Ambystoma* was taken as an index of neural disruption. As there is no definite motor horn in *Ambystoma*, the label was counted in the large neurons of the ventral part of the spinal cord and thus included both commissural and motor neurons (Herrick and Coghill, '15). It should be noted also that there are two main types of motor cells in the salamander — primary and secondary. The primary motor neurons are somewhat mediolateral in the spinal cord. They were also included in the grain counts. At various times after single or double limb amputation grain counts were done on ventral spinal neurons and cells of the brachial ganglia. These were statistically evaluated in reference to control sections from unamputated animals. In each case there were five animals per experimental group. In each animal grain counts of 25 ganglion cells were randomly made from the three brachial ganglia to arrive at a mean number of silver grains per nerve cell. At the level of the ganglia grain counts were also done on 20 ventral spinal cord cells to obtain a mean value. Mann-Whitney U tests (Siegel, '56) were run to compare the medians for animals in each group with those in other groups or with controls. Table 2 gives the data from these grain counts, corrected for background.

All injections of label were intraperitoneal and made through the tail musculature to prevent the loss of label by leakage. Ribonuclease digestion (1% for 4 hours at 40°C.) was done on slides from all groups to ascertain that the label was going into just the nucleic acid. The dipping method of autoradiography was used according to the technique of Kopriwa and Leblond ('62) with NTB-3 emulsion diluted 1:1. After development in Dektol the autoradiographs were thoroughly rinsed and stained with hematoxylin or methylene blue (1% at pH 7.0). As different amounts of label and exposure time were used at the three time intervals tested, which were at one, five, and ten days following amputation, these are also noted on table 2. To avoid counting the same cell twice only sections at 20 μ intervals were used for grain counts.

Results

One day after amputation. In the group of animals with a single forelimb removed both the ganglion cells and ventral spinal cord cells on both sides of the spinal cord showed statistically greater label incorporation ($P = 0.004$) than their respective controls when medians were evaluated by the Mann-Whitney U test. The animals of the group with both forelimbs amputated showed such great incorporation of label into ganglion and ventral spinal cord cells that it was impossible to do meaningful grain counts on them.

Five days after amputation. The ventral spinal cord cells of the group of animals with one limb amputated had returned to normal levels. However, in this group it was found that the ganglion cells on both the side of limb transection as well as the contralateral side still showed increased uridine incorporation ($P = 0.004$) as compared to their respective controls. Likewise the ganglion cells of the group with both forelimbs removed showed increased label ($P = 0.004$) as compared to their controls. It was interesting to note that the ventral spinal cord cells of the double amputation group still showed increased incorporation of label ($P = 0.004$) when compared to either the ventral spinal cord cells of the control or of the group with one forelimb removed. By ten days after amputation all label incorporation in the experimental animals had returned to control levels.

Thus, it was found that, using increased uridine incorporation as an index of neural disruption, single limb removal quickly brought about increased labeling on both sides of the spinal cord and also in the ganglia of both the side of limb removal and the contralateral side. The most unexpected result was that the contralateral ganglion cells showed increased uridine incorporation. It may be remembered that there was also slight indication of nuclear changes in the contralateral ganglia following single forelimb removal. However, the increase in uridine incorporation by these cells was of an overall increase in number of grains per cell and not due to the presence of a few cells showing greatly increased uridine incorporation, as if they had been injured only to a slight degree. The ganglion cells of the animals with one or both forelimbs removed also show continued increase of uridine incorporation at five days after limb removal. The much larger amount of cytoplasm and Nissl material (RNA) in the ganglion cells as compared to spinal cord neurons is thought to be responsible for this more prolonged reaction. It should be noted that the ventral spinal cord neurons of the group of animals with both forelimbs removed do show more prolonged incorporation of label than the

TABLE 2

Tritiated uridine incorporation into nerve cells as indicated by grain counts.[1] *A comparison of the incorporation of tritiated uridine into brachial nerve cells of Ambystoma mexicanum larvae following various amputations at one and five days after amputation. At both day 1 and 5 after amputation grain counts were done on five animals in each amputation group. The averages given each represent grain counts on 20 nerve cells for the ventral spinal cord cells and 25 nerve cells for the ganglion cells. Comparison of the incorporation of label was done statistically through the use of a Mann-Whitney U test. The results are given in Series 3.*

		Single amputation					
Control		Amputated side		Contralateral side		Double amputation	
SC Cells	Ganglia	SC Cells	Ganglia	SC Cells	Ganglia	SC Cells	Ganglia
One day post-amputation [2]							
51.61 ± 17.19	36.92 ± 9.79	76.33 ± 29.41	96.76 ± 24.63	66.16 ± 22.07	75.50 ± 20.49	>100	>100
45.80 ± 18.41	36.40 ± 15.62	79.48 ± 20.07	90.82 ± 22.29	70.12 ± 23.87	77.42 ± 18.69		
54.62 ± 15.27	37.84 ± 17.11	72.50 ± 25.18	100.12 ± 31.11	63.33 ± 16.94	78.17 ± 30.40		
50.18 ± 19.81	40.92 ± 19.64	69.49 ± 22.76	92.18 ± 27.27	64.81 ± 19.05	80.61 ± 19.34		
43.20 ± 20.22	33.42 ± 12.51	81.01 ± 19.48	95.42 ± 22.36	61.47 ± 23.77	77.00 ± 27.41		
Five days post-amputation [3]							
26.75 ± 9.31	29.72 ± 13.06	22.45 ± 8.71	60.89 ± 20.95	27.44 ± 9.01	40.00 ± 19.71	37.37 ± 9.70	54.60 ± 15.12
22.30 ± 8.62	34.82 ± 12.53	27.68 ± 8.92	63.42 ± 16.62	23.82 ± 8.46	47.92 ± 18.80	38.60 ± 14.14	56.82 ± 18.26
29.86 ± 9.05	31.44 ± 9.37	29.40 ± 12.39	59.76 ± 21.14	26.90 ± 9.23	38.41 ± 15.17	40.81 ± 17.77	62.23 ± 22.09
23.49 ± 7.18	28.68 ± 15.41	26.82 ± 9.90	56.16 ± 18.24	28.22 ± 9.84	42.06 ± 21.09	34.13 ± 13.37	53.41 ± 17.86
21.22 ± 6.98	27.96 ± 10.53	24.76 ± 8.68	64.92 ± 18.16	26.94 ± 7.71	37.43 ± 10.73	36.85 ± 11.21	60.81 ± 19.16

[1] Each number represents the average number of grains from counts done on either 20 spinal cord cells or 25 ganglion cells in a single animal.
[2] 7 μC uridine-5-^{3}H; 6 μ sections; exposed for 3 weeks; 3 hour incorporation.
[3] 5 μC uridine-5-^{3}H; 6 μ sections; exposed for 15 days; 2.5 hour incorporation.

group with one forelimb removed. This is thought to be the results of the combined effects of direct neural disruption on each side plus contralateral effects adding together to produce a chromatolytic response of longer duration. This is similar to the results obtained from the morphological study in Series 2. Ribonuclease digestion was found to remove virtually all label from the preparations.

SERIES 4

Growth relationships between double and single limb amputations in aneurogenic ambystoma larvae

Interestingly enough, under certain conditions nerves are not necessary for regeneration. In 1959 Yntema produced aneurogenic limbs by removal of the appropriate portion of the spinal cord and neural crest at an early stage in development. Amputation of the nerveless limbs was followed by typical regeneration. Whether regeneration of aneurogenic limbs results from persistence of embryonic growth factors in the non-neural tissues and/or from diminished threshold to normally present factors is unknown. This offers a unique system for the evaluation of agents in the inhibition of regeneration by the removal of a contralateral limb. If the neural connections between the 2 limbs are the determining factor, then one would not expect a growth inhibition with the removal of both forelimbs in the aneurogenic larvae.

Procedure

Ambystoma maculatum eggs obtained from Glenn Gentry were used. At the tail bud stage of embryonic development the technique of Thornton and Steen ('62) was used for producing aneurogenic limbs. After the completion of limb development, 16 animals had one forelimb removed and 12 had both forelimbs removed. These operations were made with fine iridectomy scissors through the humerus. The animals were kept in sterile Steinberg's solution with sodium sulfadiazine added to prevent bacterial infection. Throughout the experiment they were maintained in an incubator at 18°C. Both growth in length and morphogenetic stages were used to gauge the rate of regeneration. Measurements were made in this case without the use of anesthesia as the larvae are quite immobile. For statistical reasons only one of the limbs of the double forelimb amputation group was measured in analyzing the growth rate.

Results

Figure 4 gives the mean regenerate length of the limbs of the two groups beyong the original limb stump throughout limb regrowth. It can be seen that there appears to be no difference in the growth rate of the two groups. Indeed, statistical analysis by a Mann-Whitney U test at day 14 after amputation shows no statistical difference at the 5% level between the amount of linear growth achieved by the two groups at this time. The completion of digit formation was achieved by all animals by 18 days; no difference was seen between the two groups in speed of morphogenesis.

The fact that there is no growth inhibition with double forelimb removal in the aneurogenic larvae supports the idea that general body (systemic) effects are not involved.

SERIES 5

Investigation of contralateral effects on limb regeneration in the newt

The foregoing series of experiments have presented evidence of contralateral effects on the trophic role of the nerve in *Ambystoma* limb regeneration. *Ambystoma* has a large number of fibers crossing the spinal cord via the ventral commissure (Herrick and Coghill, '15) that might be significant in effecting these contralateral changes. It was therefore of some interest to determine whether such pronounced contralateral effects on growth are also present in the newt, *Notophthalmus (Triturus) viridescens,* which is used widely in regeneration studies. Therefore, a growth study was designed to determine the rate of regeneration with the removal of one or two forelimbs in various combinations in the newt. It was also designed to test whether or not the transection of only the brachial nerves on one side of the animal could bring about a slower regeneration of the forelimb of the opposite side.

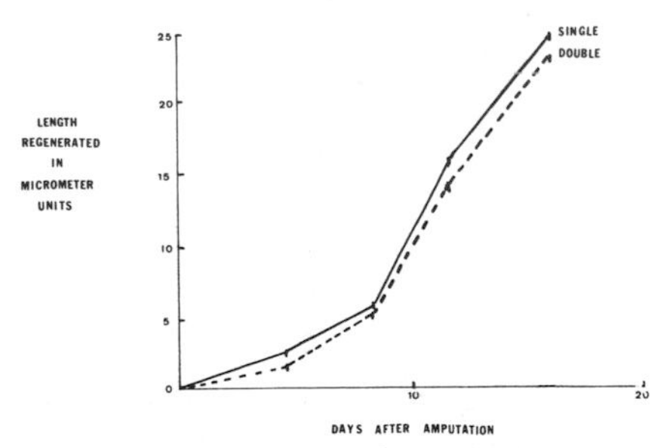

Fig. 4 A comparison between the growth rates in length of aneurogenic *Ambystoma maculatum* larvae following amputation of either one or two forelimbs. A Mann-Whitney U test run at day 14 after amputation showed no significant difference in amount of regeneration. One micrometer unit equals 0.125 mm.

Procedure

Forty adult newts were randomly divided into four groups. These were:

A. Those with only one forelimb amputated and the opposite forelimb sham denervated by cutting through the skin over the brachial plexus.

B. Those with one forelimb and one hindlimb amputated.

C. Those with only one forelimb amputated and the forelimb of the opposite side of the animal denervated by cutting through the three brachial nerves at the brachial plexus.

D. Those with both forelimbs amputated.

All amputations were made as closely as possible at the same level of the humerus or femur. The rate of regeneration was by measurement of the growth of the regenerate in length beyond the original stump length and supplemented by morphogenetic observations. For statistical purposes only one of the limbs of the double amputation group was measured. Only the forelimb of the group with both one forelimb and one hindlimb removed was measured. During the course of the experiment all animals were fed every other day on sliced beef liver and maintained at laboratory temperature of 22 ± 2°C.

Results

Figure 5 shows the mean regeneration in length of the limbs of the four groups of animals. As regeneration proceeds there appears a difference in growth rate whereby the animals in groups A and B show faster linear regeneration than the animals in groups C and D. At 45 days after amputation statistical analysis by a one-way analysis of variance and new multiple range test shows that the apparent differences are significant ($P < 0.05$). Morphogenesis in the animals of groups C and D was found to lag behind those of the other two groups by about three to four days. Neither this difference in morphogenesis nor the differences in linear growth were so clearcut as was found in *Ambystoma* larvae. A part of the reason for this probably lies in the greater variation normally seen in the regeneration rates of a group of newts as compared to a group of axolotls.

Thus, in the newt as well as in the axolotl contralateral double forelimb removal brings about a slower linear regeneration rate than single forelimb removal. The fact that neural connections may be responsible for this phenomenon is strongly supported by the fact that single forelimb amputation on the side of an animal *opposite* a severance of the brachial nerves shows slower regeneration than single forelimb amputation opposite a sham denervation. The fact that it is not the *amount* of injury or tissue removed that is the factor bringing about slower regeneration is supported by the evidence that animals with both one hindlimb and one forelimb removed show just as rapid regeneration of the forelimb as an animal with just the single forelimb removed. In summary, there exists in the newt as well as in the axolotl a localized effect on regenerative growth whereby the removal of the two contralateral limbs brings about a slower linear regeneration rate than single limb amputation. Evidence points to the involvement of the brachial nerves in this effect, as just the denervation of one forelimb of a newt will bring about a slower regeneration in the contralateral forelimb.

SERIES 6

Autoradiographic analysis of contralateral effects at an early phase of regeneration

A number of workers have reported that partial denervation of a limb brings about

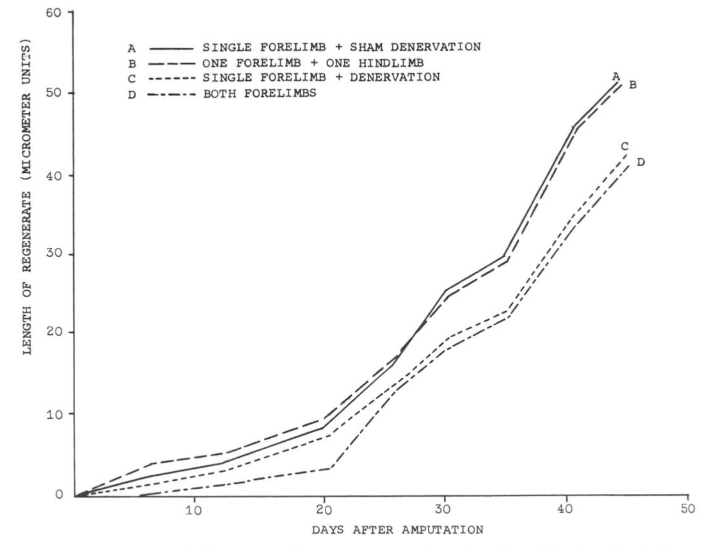

Fig. 5 A comparison of the mean linear regeneration length of limbs of adult newts following one of the following four amputation procedures: A. one forelimb removed, opposite forelimb sham denervated; B. one forelimb and one hindlimb removed; C. one forelimb removed, opposite forelimb denervated; D. both forelimbs removed. Variance analysis and a new multiple range test run at 45 days after amputation showed that groups A and B displayed significantly greater linear growth than groups C and D ($P < 0.05$). One micrometer unit equals 0.125 mm.

a decrease in regeneration rate (Schotté, '26; Weiss, '25; Karczmar, '46; Singer and Egloff, '49). This effect is apparently due to a decrease in growth during the rapid proliferation stage of limb regeneration and not through a delay in the beginning of regeneration (Singer and Egloff, '49). It may be noted from figure 5 that the difference in regeneration rates brought about by removal or denervation of a contralateral limb seems to be mainly through differences appearing in the rapid growth phase of regeneration. An early difference in the trophic effect of the limb innervation brought about by increased injury to the neurons of the brachial region might explain the latter effect on growth rate. This series was designed to test for such a difference as early as seven days after limb amputation.

Procedure

Twenty-one newts were used. These were divided into three groups based upon experimental procedure. These groups were:

A. One forelimb removed with the opposite forelimb sham denervated by breaking the skin over the brachial plexus.

B. Both forelimbs removed.

C. One forelimb removed with the opposite forelimb denervated by severance of the brachial nerves.

The animals of these groups were each given intraperitoneal injections of 4.5 μc of thymidine methyl ^{3}H (New England Nuclear, specific activity 5 C/mM) in 4.5 μl of distilled water and allowed to incorporate the label for three hours. Deoxyribonuclease (Worthington) digestion was carried out to check the specificity of labelling (5% DNase in 0.1 M tris maleate buffer at pH 6.5, incubation at 37°C for 24 hours). Following fixation in Carnoy's, decalcification was accomplished by treatment with a 10% solution of disodium versenate at pH 7.5 for 12 hours. Ten μ thick paraplast sections were cut through the longitudinal axis of the limb. Amount of incorporation of the labeled precursor into animals of the three groups was determined from autoradiographs of the longitudinal sections. By projections of the sections onto graph paper it was ascertained which was the area of the limb near the center of the limb. The number of labeled cells in this section was then counted as well as the number of labeled cells in sections at 30 and 50 μ on either side of the center section. Labeled epidermal cells were not included in these counts. These three values were then averaged for statistical evaluation. Thus, each animal was assigned an average number of labeled cells; the average number of labeled cells of the animals in every group was then compared by means of a new multiple range test and one-way analysis of variance. Only one of the forelimbs of the animals with both forelimbs removed was used for statistical evaluation.

Results

Table 3 presents graphic analysis of the data. There were found to be significantly ($P < 0.01$) more labeled cells in the control animals with only one forelimb removed (plus sham denervation of the opposite forelimb) than in the other two experimental groups. Thus, even as early as seven days after limb amputation lessened thymidine incorporation may be seen in a forelimb following the amputation or denervation of the contralateral forelimb. This indicates that the transneuronal effect on growth involves differences in mesenchymatous cell proliferation from the earliest stages.

DISCUSSION

The preceding experiments point out the presence of a decreased rate of limb regeneration in *Ambystoma mexicanum* and *Notophthalmus viridescens* following removal of a second limb but only if the

TABLE 3

A comparison of number of cells labeled with tritiated thymidine in the newt limb at seven days after removal of: A. one forelimb plus sham denervation of opposite forelimb (control); B. one forelimb plus denervation of opposite forelimb; C. both forelimbs. The sampling technique is given in Series 6.

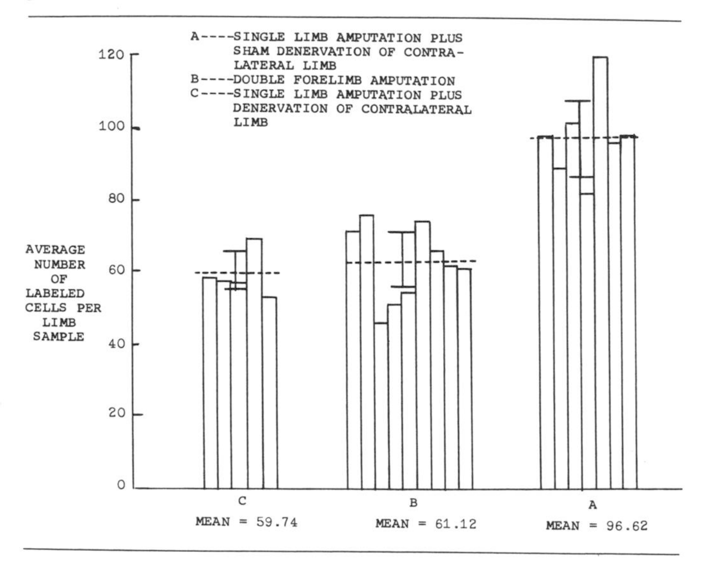

second limb removed is directly contralateral to the first. The removal of a second limb has no significant effect on regeneration rate when it is not contralaterally opposed to the first. This would deny a general systemic effect being involved. Evidence points to the presence of contralateral nerve injury as the mechanism bringing about this decrease in regeneration rate as unilateral limb removal was found to produce bilateral chromatolysis in the brachial neurons. This was seen both with morphological and autoradiographic criteria. Using these techniques, it was also found that bilateral limb extirpation brought about more profound disruption in the brachial neurons such that they showed both a greater and more prolonged chromatolysis than following unilateral limb removal. Further evidence for the involvement of just the brachial innervation of one limb of an animal on the regeneration rate of the contralateral limb is also seen in the newt where just the severance of the brachial nerves of the forelimb on one side of an animal significantly decreased the regeneration rate of the contralateral forelimb. In animals where there was no appreciable limb innervation or intervening central nervous system tissue present, i.e., the aneurogenic *Ambystoma* larvae, the removal of a contralateral limb had no effect on regeneration rate.

The fact that the additional removal of a hindlimb had no effect on the regeneration rate of a forelimb does not support the conclusion of Zeleny ('09) that additional limb or tail removal (injury) in *Ambystoma* brought about greater speed of regeneration. Reports (Zeleny, '09; Blacher, '32) of the stimulating effect of hindlimb amputation on tail regeneration in the frog tadpole remains unexplained. It should be noted that while two contralateral limbs are connected by a considerable degree of neural reflex connections, this is not the case with the tail and a hindlimb. This might explain the lack of a decreased growth rate with removal of the tail and a hindlimb but does not explain the acceleration of growth reported. For an understanding of the cause of the increased regeneration rate perhaps other work of Blacher ('32) is pertinent. In a series of experiments he found that the removal of additional pieces of fin in *Rana* tadpoles brought about accelerated regeneration rate but only if the two pieces of fin removed were close together. Perhaps the proximity of the tail to the hindlimb could explain an accelerated regeneration rate if some local growth-stimulating substance is given off by the amputation surface or regenerating structures.

Although the severity of transneuronal degeneration decreases with the maturation of the nervous system, there are still a number of reports of its occurrence in the mature nervous system — usually in the projection pathways of the brain (Hamburger, '54; Bleier, '69). There have also been reports of "transneuronal" degeneration in which an injured nerve cell brings about chromatolysis in neurons with which they have no known neural connections. One case of this was reported by Ochs et al. ('60); in this instance they noted the appearance of chromatolytic neurons in the lateral nuclear groups of both sides of the spinal cord of the cat following the crushing of a single sciatic nerve. This is in agreement with an earlier report of Barr and Hamilton ('48). Nittono ('23) observed an unexpected amount of contralateral chromatolysis in the trigeminal motor nucleus following unilateral nerve section. Since no known neural pathways could account for the bilateral chromatolysis, Nittono suggested that a diffusible substance given off from degenerating nerves might be involved. In the present study it was indicated that cellular changes occurred on both sides of the spinal cord following single forelimb amputation. Stains for degenerating nerve endings showed that fiber degeneration was present bilaterally after unilateral forelimb amputation. However, it is still unresolved whether the degree of "transneuronal" changes seen which included effects on the ganglion *contralateral* to the limb removed can completely be explained either by anterograde or retrograde transneuronal trauma. Perhaps there is some diffusible substance or breakdown product given off by the directly injured nerve cells that diffuses across the spinal cord to bring about chromatolysis in contiguous cell bodies and opposite ganglia.

It should also be pointed out that two indices of chromatolysis chosen — increased uridine incorporation and nuclear chromatin changes — may be more sensitive indicators of changes in the neuronal integrity than the classic Nissl breakdown. These techniques could be useful as sensitive tools in other studies on transneuronal effects.

It has been suggested (Nandy, '68; Ducker et al., '69) that in chromatolysis the nerve cell body has to increase its proteosynthesis first to survive and secondly to provide the metabolic environment necessary to rebuild damaged parts, such as severed axons. Experimental support for this contention comes from the work of Cavanaugh ('51) who found that the surviving cell bodies of capped intercostal nerves of the rat remained in a permanent state of chromatolysis; without neural outgrowth the cell body did not leave the chromatolytic state. Bodian and Mellors ('45) suggest that the phases of neuronal repair following nerve section are related to the lengthening of the regenerating axon. If with the greater degree of neuronal injury, more cellular energy must be devoted to repair of the perikaryon, material for fiber regeneration or maintenance of the end organ may be delayed. A delay in effective neural regeneration could explain a decrease in rate of limb regeneration; it was shown many years ago that partial denervation of a limb, by cutting various combinations of the three brachial nerves that supply the limb, will bring about a decreased regeneration rate. In a series of experiments Singer ('52) calculated the number of nerve fibers at various levels of the limb necessary to obtain limb regeneration. Below a certain "threshold" number of fibers per unit area of limb tissue no regenerative response ensued. In the range of innervation between the normal supply and threshold level, partially denervated limbs may be obtained. Schotté ('26) and Weiss ('25) reported that in *Triton* there is an additive effect of nerves on blastemal growth such that they found a graded reduction in limb innervation brought about a graded decrease in rate of regeneration. Karczmar ('46) also supported the idea that the response to the amount of innervation was a graded one, with a delay in the onset of regeneration coming about in limbs with a reduced nerve supply. Singer and Egloff ('49) found that partially denervated adult limbs do regenerate at a slower rate but that the rate of regeneration is not a graded response to amount of innervation. They did not believe that delayed onset of regeneration was the reason for diminished regeneration rate, but instead suggested that it was due more to a slowing down in the successive stages of growth. Similar results are felt to pertain to the present experiments, where, although there were differences in thymidine incorporation as early as seven days after amputation, the greatest differences in growth appear in later stages (see figs. 1, 5).

Using another approach to the question of whether there was an additive effect of nerves on regeneration rate, Shuraleff ('68) undertook experiments on the effect of nerve augmentation in *Ambystoma* larvae. He found that the presence of an augmented nerve supply in the limb was correlated with accelerated growth and limb morphogenesis but not with the final size of the regenerate. "Superinnervation" was not related to the time of appearance of mesenchymatous cells at the wound surface, but did bring about increased mesenchymatous cell proliferation. The influence of nervous tissue on mitosis was also demonstrated by Overton ('50, '55). She found that *Ambystoma* spinal cord tissue grafted homoplastically to the dorsal fin stimulated growth of the fin. She was able to determine that the "active principle" was soluble in water and relatively resistant to alcohol, ether, and cold.

Recently Lebowitz and Singer ('70) have provided additional evidence for the chemical nature of the neurotrophic effect in limb regeneration. They found that homogenates or supernatants of brachial nerve could restore about 60% of denervation-induced inhibition of protein synthesis when microinfused into early newt blastemata. Singer ('60, '65) has postulated axonal flow as a possible mechanism for delivery of a neurotrophic substance to the regenerating limb. A delay in the delivery of such a material to the limb because of prolonged nerve cell disruption or slower nerve fiber regrowth could then be a possible explanation of the contralateral effects on growth reported here.

ACKNOWLEDGMENTS

This investigation was conducted as part of a dissertation submitted to Michigan State University in partial fulfillment of the requirements for the Ph.D., August 1970. It was supported by a National Institute of Health traineeship (HD-00135) and by NIH grant NBO-04128 administered by Dr. C. S. Thornton. I would like to express sincere thanks and appreciation to Dr. Thornton for his excellent guidance in the course of this investigation.

LITERATURE CITED

Barr, J., and J. Hamilton 1948 A quantitative study of certain morphological changes in spinal motor neurons during axon reaction. J. Comp. Neur. *89:* 93-121.

Berube, G., M. Powers and G. Clark 1966 The gallocyonin-chromealum stain: influence of methods of preparation. Stain Technology, *41:* 73-81.

Blacher, L., A. Irichimowitsch, L. Liosner and M. Woronzowa 1932 Uber den Einfluss des Regenerationprozesses eines Teiles des Organismus auf die Geschwindigkeit der Regeneration eines anderen Teiles. Arch. Entwicklung., *127:* 370-386.

Bleier, R. 1969 Transsynaptic degeneration in young and adult rabbits. Brain Res., *15:* 365-385.

Bodian, D., and R. Mellors 1945 The regenerative cycle of motoneurons with special reference to phosphatase activity. J. Exp. Med., *81:* 469-487.

Causey, G., and C. Strattman 1959 Changes in the DNA content of chromatolytic rabbit neurones. J. Anat., *93:* 341-348.

Cavanaugh, M. 1951 Quantitative effects of the peripheral innervation area on nerves and spinal ganglion cells. J. Comp. Neur., *94:* 181-196.

Clemente, C. 1964 Regeneration in the Vertebrate central nervous system. Ann. Rev. Neurobiol., *6:* 258-293.

Cole, M. 1968 Retrograde degeneration. In The Structure and Function of Nervous Tissue. Vol. 1. G. Bourne, ed. John Wiley and Sons, New York.

Detwiler, S. 1936 Neuroembryology, an Experimental Study. The Macmillan Co., New York.

Dresden, M. 1969 Denervation effects on newt limb regeneration: DNA, RNA, and protein synthesis. Dev. Biol., *19:* 311-321.

Ducker, T., L. Kempe and G. Hayes 1969 The metabolic background for peripheral nerve surgery. J. Neurosurgery, *30:* 270-280.

Fink, R., and L. Heimer 1967 Two methods for selective silver impregnation of degenerating axons and their synaptic endings in the central nervous system. Brain Res., *4:* 369-374.

Gray, F., and R. Guillery 1966 Synaptic morphology in the normal and degenerating nervous system. Int. Rev. Cytol., *19:* 111-182.

Haddad, A., S. Iucif and A. Cruz 1969 Synthesis of RNA in neurons of the hypoglossal nerve nucleus of mice after section of the axon in mice. J. Neurochem., *16:* 865-868.

Hamburger, V. 1954 Trends in experimental embryology. In: Proc. 1st Internat. neurochem. Symp. Waelsch, ed. Academy Press, New York.

Herrick, C., and G. Coghill 1915 The development of reflex mechanisms in *Ambystoma.* J. Comp. Neur., *25:* 65-85.

Hiller, S. 1926 As cited by L. Blacher, A. Irichimowitsch, L. Liosner and M. Woronzowa 1932 Uber den Einfluss des Regenerationprozesses eines teiles des Organismus auf die Geschwindigkeit der Regeneration eines Teiles. Arch. Entwicklung., *127:* 370-386.

Huxley, J., and G. De Beer 1934 Elements of Experimental Embryology. Cambridge University Press.

Issekutz-Wolsky, M., and L. Fogarty 1962 Effect of regeneration blastema autotransplantation on subsequent regeneration in *Triturus (Diemictylus) viridescens.* Nature, *195:* 621-622.

Karczmar, A. 1946 The role of amputation and nerve resection in the regression of urodele larvae. J. Exp. Zool., *103:* 401-427.

Kasten, F., and V. Burton 1959 A modified Schiff's solution. Stain Technology, *34:* 289-293.

Kopriwa, B., and C. Leblond 1962 Improvements in the coating technique of radioautography. J. Histochem. and Cytochem., *10:* 269-284.

Lebowitz, P., and M. Singer 1970 Neurotrophic control of protein synthesis in the regenerating limb of the newt, *Triturus.* Nature, *225:* 824-827.

Lentz, T. 1967 Fine structure of sensory ganglion cells during limb regeneration of the newt *Triturus.* J. Comp. Neur., *131:* 301-322.

Lison, S. 1962 C. R. Acad. Sci. (Paris), *254:* 2239-2243.

Morgan, T. 1906 Physiology of regeneration. J. Exp. Zool., *3:* 457-500.

Nandy, K. 1968 Histochemical study on chromatolytic neurons. Arch. Neurol., *18:* 425-434.

Needham, A. 1952 Regeneration and Wound Healing. Meuthen and Co. Ltd., London.

Nittono, K. 1923 On bilateral effects from unilateral section of the *nervus trigeminus* in the albino rat. J. Comp. Neur., *35:* 133-161.

Noeske, K. 1969 Diskrepanzen von Feulgen-Wert und DNS-Gehalt. Histochemie, *20:* 322-327.

Overton, J. 1950 Mitotic stimulation of amphibian epidermis by underlying grafts of central nervous tissue. J. Exp. Zool., *115:* 521-561.

——— 1955 Mitotic responses in amphibian epidermis to feeding and grafting. J. Exp. Zool., *130:* 433-483.

Ochs, S., R. Kachmann and W. Demyer 1960 Axoplasmic flow rates during nerve regeneration. Exp. Neurol., *2:* 627-637.

Schotté, O. 1926 Systeme nerveux et regeneration chez le triton. Rev. suisse Zool., *87:* 297-313.

Shuraleff, N. 1968 Doctoral thesis, Michigan State University, East Lansing, Michigan.

Siegel, S. 1956 Nonparametric Statistics. McGraw-Hill, New York.
Singer, M. 1952 The influence of the nerve in regeneration of the amphibian extremity. Quart. Rev. Biol., *27:* 169-200.
——— 1959 The influence of nerves on regeneration. In: Regeneration in Vertebrates. C. S. Thornton, ed. University of Chicago Press, Chicago.
——— 1960 Nervous mechanisms in the regeneration of body parts in vertebrates. In: 18th Symp. of Growth and Development. Ronald Press, New York.
——— 1965 A theory of the trophic nervous control of amphibian limb regeneration, including a reevaluation of quantitative nerve requirements. In: Regeneration in Animals and Related Problems. Kiortsis and Trampusch, eds. North-Holland Publishing Co., Amsterdam.
Singer, M., and F. Egloff, 1949 The nervous system and regeneration of the forelimb of adult *Triturus.* VIII. The effect of limited nerve quantities on regeneration. J. Exp. Zool., *111:* 295-315.
Skowron, S., K. Rzehak and K. Maron 1963 The effect of blastemal tissue on regeneration. Folia Biol. (Krokow), *11:* 259-267.
Thornton, C., and T. Steen 1962 Eccentric blastema formation in aneurogenic limbs of *Ambystoma* larvae following epidermal cap deviation. Dev. Biol., *5:* 328-343.
Twitty, V., and L. Delanney 1939 Size regulation and regeneration in salamander larvae under complete starvation. J. Exp. Zool., *81:* 399-414.
van Gehuchten, A. 1898 La moelle epiniere des larves des betraciens. Arch. de Biol., *15:* 599-620.
Watson, W. 1965 An autoradiographic study on the incorporation of nucleic acid precursors by neurones and glia during nerve regeneration. J. Physiol., *180:* 741-753.
Weber, M., and K. Maron 1965 The effect of blastemal tissue on regeneration. II. Influence of blastemal cell fractions on regenerative growth. Folia Biol., *13:* 383-396.
Weiss, P. 1925 Abhangigkeit der Regeneration entwickelter Amphienextremitaten von Nervensystem. Arch. Entwicklung. *104:* 317-358.
Wyburn-Mason, R. 1950 Trophic Nerves. Kimpton, London.
Yntema, C. 1959 Regeneration in sparsely innervated and aneurogenic forelimb of Amblystoma larvae. J. Exp. Zool., *140:* 101-124.
Zeleny, C. 1909 Degree of injury and its effect upon regeneration rate. J. Exp. Zool., *7:* 513-561.

Fine Structure of Sensory Ganglion Cells During Limb Regeneration of the Newt *Triturus* [1]

THOMAS L. LENTZ

Limb regeneration of the newt is dependent upon the nervous system and is one of the most dramatic examples of the trophic function of nerves. The sensory supply to the limb is of major importance because it will sustain regeneration in the absence of motor and sympathetic components (Singer, '43). All nerves appear to have a trophic effect but normally only the sensory supply meets the necessary quantitative conditions: one-third to one-half of the total number of nerve fibers must be present in the limb for regeneration to occur (Singer, '46). The trophic effect arises within the sensory cell bodies because a sensory supply, isolated from its central connections by section of the dorsal roots, will sustain regeneration (Sidman and Singer, '51). Regeneration does not occur, however, when the ganglia are destroyed (Singer, '43). Sensory ganglia are also capable of sustaining regeneration when implanted into denervated limbs (Kamrin and Singer, '59).

The present experiments were undertaken to study the mechanism by which nerves exert a trophic effect during regeneration. The fine structural events occurring in the peripheral nerves during limb regeneration have been described (Lentz, '67). In this communication, the fine structure of sensory ganglion cells during limb regeneration of the newt is presented.

MATERIALS AND METHODS

Operations were performed on adult newts, *Triturus viridescens*, anesthetized in MS-222 (tricaine methanesulfonate, Sandoz Pharmaceuticals, Hanover, New Jersey). The limbs were amputated through the lower third of the upper forelimb. The protruding end of the humerus was removed and the animals placed on damp paper in covered fingerbowls. They were fed chopped beef liver once a week and allowed to regenerate for 1, 3, 7, 14, 21, or 35 days. The latter time corresponds to the emergence of digits in the regenerating limb. At the end of each period, sensory ganglia were removed for fixation. For controls, the ganglia of normal newts or the

[1] This work was supported by a grant from the National Cancer Institute (TICA-5055), National Institutes of Health, United States Public Health Service.

ganglia on the unoperated side of experimental animals were removed.

Spinal nerves 3, 4, and 5 supply the forelimb (Singer, '42). Although the ganglia appear to be structurally similar, the fourth dorsal root ganglion was used most frequently. The ganglion was removed by making an incision parallel to the vertebral column and down to the transverse process of the vertebra, chipping away the transverse process and portion of the rib to expose the ganglion, and cutting the roots and spinal nerve to free the ganglion (Singer, '42, '43).

After removal, the ganglia were placed immediately in cold 2–5% glutaraldehyde in 0.05 M cacodylate buffer (pH 7.2) for one hour. The tissues were rinsed overnight in cold buffer and refixed for one hour in cold 1% osmium tetroxide buffered with veronal acetate (pH 7.2) and containing 0.2 M sucrose. The ganglia were dehydrated rapidly in cold ethanol and embedded in Maraglas (Freeman and Spurlock, '62). Thin sections were cut on a Porter-Blum microtome, stained with lead hydroxide (Feldman, '62), and examined with an RCA EMU 3F electron microscope. For orientation with the light microscope, 1-μ plastic sections were cut and stained with 0.1% toluidine blue.

OBSERVATIONS

Normal sensory ganglion. Normal ganglion cells are round or oval and have a large paracentrally situated nucleus (fig. 1). The nucleus contains few indentations and is bounded by an envelope perforated by a few pores. The nucleoplasm is occupied by dense granules, 300–400A in diameter, and smaller particles and filamentous material. This material is usually evenly dispersed but may occur in small clumps. Each nucleus has one or two nucleoli that may be up to 3 μ in diameter (fig. 2). The large nucleoli are composed of fine textured fibrillar material and small granules. The fibrillar material occurs in one or two large round clumps (pars amorpha) and within large anastomosing strands (nucleolonema). The granules are about 150A in diameter and are present in all regions of the nucleolus except the pars amorpha. Accessory nucleoli, about 0.6 μ in diameter, are situated near the large nucleoli and are composed of a tightly packed mass of fine fibrillar and granular material (fig. 2).

Round and oval mitochondria, 0.2 μ in diameter, occur throughout the cytoplasm (fig. 3). These mitochondria contain a few short cristae. Cristae are more numerous in larger mitochondria (0.5 μ) and extend transversely across the organelle. The low density matrix contains two or three opaque granules.

The endoplasmic reticulum is composed primarily of rough-surfaced cisternae. Short, irregularly shaped profiles of cisternae are scattered throughout the cell but are often collected into zones (Nissl bodies) relatively free of other organelles (figs. 1, 3). The cisternae may be slightly dilated and contain fine-textured material of medium density. Ribosomes are attached to the cisternae and also occur free in the cytoplasm, especially in the Nissl zones. The cisternae of endoplasmic reticulum are only partially covered by ribosomes, short regions of the membrane being devoid of attached ribosomes. Cisternae of smooth surfaced endoplasmic reticulum, more tubular and of smaller diameter than rough cisternae, are connected to the rough-surfaced cisternae (fig. 1). The channels of smooth endoplasmic reticulum, not numerous in the perikaryon, occur in the axon and extend into the axon hillock (fig. 1).

Large round, oval, or irregular membrane-bounded bodies 0.5 μ in diameter occur in the ganglion cells (fig. 3). The content of these structures differs. In some, particularly those with a regular outline, the contents are finely granular and of uniform moderate density (fig. 3). These structures are probably lysosomes. Smaller vacuoles (0.1–0.2 μ) but with identical contents occur near some Golgi complexes (fig. 3). The contents of other bodies, lipofuscin pigment granules, are heterogeneous with opaque, granular, or lucent regions (fig. 3). Some also contain parallel arrays of membranes. A few are composed almost entirely of homogeneous opaque material.

The hyaloplasm is of low density (figs. 1, 3). Immediately external to the nuclear envelope, the cytoplasm contains fewer organelles than elsewhere. Microtubules course circumferentially around the nucleus in this region splaying out into the cytoplasm especially in the region facing the axon hillock (fig. 3). Several Golgi complexes are situated parallel to the nuclear envelope and on the periphery of the perinuclear zone of cytoplasm (fig. 3). Others occur throughout the cell. These organelles are composed of elongated membranous cisternae and small vesicles. The cisternae facing the nucleus are shorter and more dilated than those facing the cytoplasm which are more flattened and elongated. Vesicles are most numerous at the ends of the lamellae and occur along the face of the Golgi apparatus on the side facing the cytoplasm. Most of the vesicles are 300–400A in diameter with lucent or moderately dense contents. Some vesicles are larger (500–700A) and contain a dense core (fig. 3). Vesicles are most numerous in the vicinity of the Golgi zone but occur singly in other regions of the cell.

The axon hillock contains relatively few organelles (fig. 1). A few small mitochondria, membranous profiles, ribosomes, or vesicles may occur in this region. Microtubules extend from the perikaryon into the axon. The proximal region of the axon is similar in content to the hillock but contains more tubular profiles of smooth-surfaced endoplasmic reticulum. These structures are 250–300A in diameter, contain material of low density, and pursue a wavy course in the longitudinal axis of the fiber. Those extending into the perikaryon appear to connect with cisternae of rough-surfaced endoplasmic reticulum (fig. 1).

Sensory ganglion during limb regeneration. It is difficult to establish a sequence of changes in sensory ganglion cells following amputation of the limb because the changes differed in intensity in different cells. Many cells showed no observable differences from controls. The changes are considered in approximately the order of their appearance although there was considerable overlap and variation.

One of the earliest changes was an increase in the number and size of lipofuscin pigment bodies (fig. 4). Three days after amputation, they are extremely common in some cells but subsequently appear to decrease in number. At this time, these structures contain several round or oval masses of opaque material and a single large round droplet with an opaque periphery and a lucent core. Membranous lamellae also are more numerous than normal within these granules. Intermediate stages between lipofuscin pigment bodies and lysosomes are characterized by the presence of a small zone of opaque material or a few membranous lamellae occurring in the finely granular material comprising normal lysosomes (fig. 4).

The appearance of giant mitochondria is perhaps the most striking change in the spinal ganglion cells during limb regeneration (figs. 5–7). Giant mitochondria first appear at one day following amputation. Cells containing these mitochondria are most numerous from three to 21 days and decrease thereafter so that only a few occur by 35 days. At the earlier times, especially, all gradations in size from the small normal appearing mitochondria to giant mitochondria are present (fig. 5). Images suggesting the fusion of the small mitochondria to produce larger ones were not observed although small mitochondria are not nearly as abundant as normal in cells containing giant mitochondria.

Giant mitochondria may have a diameter of 2 μ, ten times that of many normal mitochondria. The cristae arise as infoldings of the inner mitochondrial membrane in the usual manner and pursue an irregular course toward the center of the organelle. In the large mitochondria, most of the cristae are tubular in cross section (fig. 7). Some mitochondria contain stacks of two or three central cristae arranged in parallel. The matrix is composed of finely granular, moderately dense material. A large number of opaque granules occur in the matrix. In some mitochondria, the cristae and opaque granules are lacking in a large central zone composed of granular material resembling and continuous with the matrix between the cristae (fig. 7).

Associated with the giant mitochondria are small Golgi complexes different from those occurring in normal cells and in the perinuclear zone during regeneration (figs. 6, 7). The latter do not appear to differ significantly from normal. Small vesicles, dense core vesicles, and vacuoles containing granular or opaque material are associated with the perinuclear Golgi complexes (fig. 8). Those occurring adjacent to giant mitochondria have shorter lamellae and fewer small vesicles. Dense membrane-bounded granules are sometimes present in relation to these Golgi complexes (fig. 7). These granules are larger (800–1100A) than the dense core vesicles (500–700A) occurring normally. They also lack the fine granular appearance of lysosomes and their precursors. A few large granules are situated in other regions of the cytoplasm. Golgi complexes containing dense granules were observed only in those cells containing giant mitochondria.

A second type of mitochondrial change occurs somewhat later (14–21 days) in different cells than those containing large mitochondria. This change consists of the appearance of a large cluster of mitochondria between the perinuclear halo of cytoplasm and the cell border (fig. 8). Two hundred fifty to three hundred mitochondria have been observed in a single section through the cluster. Some of the mitochondria comprising the cluster are slightly larger than normal but generally they resemble those of the normal cell. Very little space occurs btween the mitochondria and this is occupied by ribosomes but few other organelles.

From three to seven days after amputation, the cisternae of endoplasmic reticulum are more dispersed and less often occur in Nissl bodies (figs. 4, 9). The cisternae are slightly dilated, contain material of low density, and have fewer attached ribosomes (fig. 9). Cytoplasmic ribosomes, on the other hand, appear more numerous and may occur in small clusters (fig. 9). At 7 to 35 days, the endoplasmic reticulum becomes disorganized indicating a chromatolytic reaction (Andres, '61; Smith, '61; Bodian, '64).

In chromatolytic cells (fig. 10), the amount of rough-surfaced endoplasmic reticulum is reduced and confined to small, usually peripheral, clumps. Free ribosomes also appear to be diminished per unit area of cytoplasm. The regions of cytoplasm previously occupied by Nissl substance are filled with expanses of flocculent material and dense granules resembling particulate glycogen. Mitochondria and lipofuscin granules are also excluded from these regions and appear to be diminished in number in cells showing marked chromatolysis. The Golgi apparatus, on the other hand, does not show a reduction in size. Giant mitochondria or increased numbers of mitochondria were not observed in cells showing extreme chromatolysis.

When chromatolysis of Nissl substance occurs, the smooth-surfaced endoplasmic reticulum is likewise diminished in the perikaryon. In the proximal region of the axon, however, the smooth channels of endoplasmic reticulum and microtubules are more numerous, the channels containing moderately dense material (fig. 11). The increase in number of these structures seems to be limited to the axon because more were not noted in the perikaryon or hillock region.

Nuclear changes are most apparent in cells showing chromatolysis but also occur to a lesser extent in other cells. The nucleus sometimes appears more eccentrically situated and contains shallow indentations on its surface (fig. 10). Within the nucleus, the granular and fibrillar material becomes more aggregated occurring in small clumps (fig. 10). The latter occur throughout the nucleus and adjacent to the nuclear envelope. The nucleolus is often reduced in size and consists mainly of fibrillar material indicating a loss of its granular component.

The pattern of variation following limb amputation is complex because all of the changes in the structure of sensory ganglion cells do not occur simultaneously, to the same degree, or in the same cell, and many cells appear normal. Dense bodies (lipofuscin pigment granules) are numerous in cells that otherwise appear normal and are also abundant in a few cells containing giant mitochondria, especially at the earlier times (1 to 3 days). Some cells with giant mitochondria contain normal amounts of rough-surfaced endoplasmic reticulum or a greater density of free ribosomes. A greater number contain few dense bodies and endoplasmic reticulum confined to small peripheral zones. The small Golgi complexes with dense granules occur in cells with giant mitochondria and are situated adjacent to the mitochondria. Only a few cells show marked chromatolysis and these cells contain fewer lipofuscin pigment granules and mitochondria than normal. A less extensive chromatolytic reaction or changes in the endoplasmic reticulum occur in a greater number of cells including some of those showing other changes.

DISCUSSION

Several changes were observed in the structure of ganglion cells following amputation of the limb. The changes differed qualitatively and quantitatively in different cells. The different responses may reflect the proportion of the nerve cell lost by transection of the axons (Bodian and Mellors, '45). Cells whose axons terminated proximal to the level of amputation would not be expected to show any changes. Less change probably occurs in cells with axons that end a short distance beyond the level of amputation and the most marked changes may occur in cells with the longest axons. The regenerative capacity of the cell and associated structural changes may also be affected by the extent of initial injury. For these reasons, it is difficult to establish a sequence of events associated with retrograde chromatolysis following axon section, regeneration of the axon, or the trophic effects on the peripheral tissues.

A correlation of the changes in the perikaryon with those in the peripheral nerves during regeneration might provide some information concerning the trophic mechanism. The major differences in the sensory ganglion cells following limb amputation are an increase in the number of lipofuscin granules, hypertrophy and hyperplasia of mitochondria, occurrence of Golgi complexes with dense granules, and chromatolytic changes in the endoplasmic reticulum. In the peripheral nerves of the regenerating limb, the majority of which are sensory, three conspicuous differences in comparison to normal nerves were noted. These are the appearance of membrane-bounded granules, 1000A in diameter, an increase in the number of smooth-surfaced channels of endoplasmic reticulum, and a greater number of microtubules (Lentz, '67).

Dense membrane-bounded granules appear in both the peripheral nerves and perikaryon during limb regeneration. In the perikaryon, they occur in association with small Golgi complexes scattered among the giant mitochondria. These granules are larger than the dense core vesicles that are most numerous in the large perinuclear Golgi complexes. The small Golgi regions appearing during regeneration could be the source of the granules in the peripheral fibers. It appears, however, that fewer dense granules occur in the cell body in comparison to the peripheral fibers where granules are extremely abundant. It is possible that granules are transported down the axon as rapidly as they are formed or that smaller vesicles originating in the Golgi apparatus progressively accumulate material during axonal transport. Regarding the latter suggestion, it is of interest that the material comprising the granules is nearly identical in texture and density to that occurring within the cisternae of the peripheral tubular channels. If this similarity is more than coincidental, it is possible that vesicles are capable of incorporating material synthesized or stored within the tubular channels of endoplasmic reticulum or that the vesicles originate directly from the tubular channels. In some cases, large vesicles containing dense material are continuous with the tubular channels as though in the process of budding off (Lentz, '67). These findings indicate that while some granules may originate in the perikaryon, the majority are formed in the peripheral regenerating axons.

Both the proliferation of smooth channels and increase in microtubules are largely axonal phenomena, especially in the distal regenerating portions of the axons. The numbers of these structures in the perikaryon are comparable normally and during regeneration except in chromatolytic cells where they are reduced. The membranous channels in the proximal regions of the axon contain dense material within their cisternae during regeneration in contrast to normal. Smooth channels extend into the axon hillock toward and connecting with the cisternae of rough-surfaced endoplasmic reticulum. Normally, some dense material occurs within the cisternae of rough endoplasmic reticulum while the contents of the smooth channels are of low density. Some cells show dissolution of rough endoplasmic reticulum or slight swelling of the cisternae but additional dense material within the cisternae was not observed. This observation indicates that the dense material filling the peripheral smooth tubular channels during regeneration is synthesized or accumulated locally by these smooth cisternae although formation of material by the rough-surfaced endoplasmic reticulum and direct transfer to the connecting smooth channels could occur also.

Increase in mitochondrial number has been noted in neurons following axonal section (Hartmann, '54; Hudson et al., '61; Smith, '61). Hudson et al. ('61) also noted that the relative volume of individual mitochondria was greater after operation but this was attributed to mitochondrial swelling. Smith ('61) observed the appearance of a perinuclear cluster of sometimes hypertrophied mitochondria in dorsal root ganglia after section of the sciatic nerve. Mitochondrial enlargement very similar to that observed in the present study occurred in cultures of rat dorsal root ganglia following x-irradiation (Masurovsky et al., '67).

During limb regeneration, a few cells contain greatly increased numbers of mitochondria while a larger number of cells contain giant mitochondria with a diameter up to ten times that of normal. The alterations in mitochondrial structure may represent degenerative changes following nerve injury and associated with derangement of mitochondrial function (Masurovsky et al. ('67). Although mitochondrial enlargement occurs following neuronal injury, it may not represent primarily a degenerative change but instead indicate a change or reorganization of the energy-yielding enzymatic assemblies occurring in response to the different functional demands placed on the nerve cell after nerve section or injury. For example, mitochondrial hypertrophy and hyperplasia, if these increase the respiratory and phosphorylative units, may indicate increased oxidative activity in response to the increased energy requirements for the high rate of protein and RNA metabolism occurring during regeneration. In support of this suggestion is the finding of Hamberger and Sjostrand ('66) that respiratory enzyme activities are increased in the hypoglossal nucleus during nerve regeneration.

Lipofuscin pigment granules are increased in number, size, and complexity of internal structure following amputation of the limb. The pigment bodies appear to be derived from lysosomes because of the presence of intermediate stages. The common occurrence of a large dense droplet with a lucent core in enlarging granules supports the suggestion of Samorajski et al. ('65) that transformation of lipofuscin bodies represents an increase in their lipid moiety. The occurrence of membranous lamellae within these bodies may indicate the presence of phospholipids. Regarding the functional significance of the changes in lipofuscin granules during regeneration, a correlation has been observed between lipofuscin deposition and oxidative metabolic activity (Friede, '62).

The endoplasmic reticulum undergoes changes following amputation of the limb progressing in a few cells to almost complete loss of rough-surfaced cisternae. Prior to this stage, the proportion of free ribosomes to attached ribosomes increases. Cytoplasmic ribosomes are more numerous and some of these may originate by detachment from the cisternae of rough endoplasmic reticulum. Others may be derived from the nucleolus because the granular component of the nucleolus, thought to consist of ribosomal RNA (Birnstiel et al., '63; Stevens, '64; Jones, '65), diminishes during regeneration. Brattgård et al. ('58) also observed that the state of RNA aggregation changes after nerve section resulting in dispersion of RNA particles while the total amount of RNA is constant. They suggested that these changes represent a transition from an active to a more active form of RNA which initiates the protein synthesis occurring during outgrowth of nerve fibers. The present observations support this conclusion because the free ribosomes occur in clusters resembling polysomes which are associated with synthesis of proteins (Warner et al., '63) that may be for endogenous use, for example, soluble protein of the cytoplasmic matrix (Kuff et al., '66). Chromatolysis following axon section, therefore, appears to represent a reorganization of the synthetic machinery of the cell occurring in response to the different functional demands occurring during regeneration of the nerve fiber and possibly trophic activity.

Limb regeneration in newts depends on the nervous system and some of the structural changes occurring in nerve cells during regeneration might be associated with the trophic action of nerves. In the peripheral regenerating nerves, it was felt that large membrane-bounded granules were most likely the morphological counterpart of a trophic substance (Lentz, '67) because of their resemblance to neurosecretory granules and the association of neurosecretion with regeneration in other animals (Lender and Klein, '61; Clark et al., '62; Lentz, '65a,b). The present studies indicate that some of these granules may originate in the nerve cell body although more may be formed peripherally. The nerve cell body shows other changes during regeneration including mitochondrial changes indicative of increased metabolic activity and reorganization of the protein synthetic machinery. While some of these changes are undoubtedly associated with regeneration of the axon, they may also be related to the formation of a trophic substance.

LITERATURE CITED

Andres, K. H. 1961 Untersuchungen über morphologische Veränderungen in Spinalganglien während der retrograden Degeneration. Z. Zellforsch., *55:* 49–79.

Birnstiel, M. L., M. I. H. Chipchase and B. B. Hyde 1963 The nucleolus, a source of ribosomes. Biochim. Biophys. Acta, *76:* 454–462.

Bodian, D. 1964 An electron-microscopic study of the monkey spinal cord. I. Fine structure of normal motor column. II. Effects of retrograde chromatolysis. III. Cytologic effects of mild and virulent poliovirus infection. Bull. Johns Hopk. Hosp., *114:* 13–119.

Bodian, D., and R. C. Mellors 1945 The regenerative cycle of motoneurons, with special reference to phosphatase activity. J. Exp. Med., *81:* 469–487.

Brattgård, S. -O., J. E. Edström and H. Hydén 1958 The productive capacity of the neuron in retrograde reaction. Exp. Cell Res., Suppl., *5:* 185–200.

Clark, R. B., M. E. Clark and R. J. G. Ruston 1962 The endocrinology of regeneration in some errant Polychaetes. In: Neurosecretion. Heller, H. and R. B. Clark, eds., Academic Press, New York, 275–286.

Feldman, D. G. 1962 A method of staining thin sections with lead hydroxide for precipitate-free sections. J. Cell Biol., *15:* 592–595.

Freeman, J. A., and B. O. Spurlock 1962 A new epoxy embedment for electron microscopy. J. Cell Biol., *13:* 437–443.

Friede, R. L. 1962 The relation of the formation of lipofuscin to the distribution of oxidative enzymes in the human brain. Acta Neuropath., *2:* 113–125.

Hamberger, A., and J. Sjostrand 1966 Respiratory enzyme activities in neurons and glial cells of the hypoglossal nucleus during nerve regeneration. Acta Physiol. Scand., *67:* 76–88.

Hartmann, J. F. 1954 Electron microscopy of motor nerve cells following section of axones. Anat. Rec., *118:* 19–34.

Hudson, G., A. Lazarow and J. F. Hartmann 1961 A quantitative electron microscopic study of mitochondria in motor neurones following axonal section. Exp. Cell Res., *24:* 440–456.

Jones, K. W. 1965 The role of the nucleolus in the formation of ribosomes. J. Ultrastruct. Res., *13:* 257–262.

Kamrin, A. A., and M. Singer 1959 The growth influence of spinal ganglia implanted into the denervated forelimb regenerate of the newt, *Triturus*. J. Morph., *104:* 415–440.

Kuff, E. L., W. C. Hymer, E. Shelton and N. E. Roberts 1966 The in vivo protein synthetic activities of free versus membrane-bound ribonucleoprotein in the plasma-cell tumor of the mouse. J. Cell Biol., *29:* 63–75.

Lender, T., and N. Klein 1961 Mise en évidence de cellules sécrétrices dans le cerveau de la Planaire *Polycelis nigra*. Variation de leur nombre au cours de la régénération posterieure. C. R. Acad. Sci. (Paris), *253:* 331–333.

Lentz, T. L. 1965a Fine structural changes in the nervous system of the regenerating hydra. J. Exp. Zool., *159:* 181–194.

——— 1965b Hydra: Induction of supernumerary heads by isolated neurosecretory granules. Science, *150:* 633–635.

——— 1967 Fine structure of nerves in the regenerating limb of the newt *Triturus*. Am. J. Anat., in press.

Masurovsky, E. B., M. B. Bunge and R. P. Bunge 1967 Cytological studies of organotypic cultures of rat dorsal root ganglia following X-irradiation in vitro. I. Changes in neurons and satellite cells. J. Cell Biol., *32:* 467–496.

Samorajski, T., J. M. Ordy and J. R. Keefe 1965 The fine structure of lipofuscin age pigment in the nervous system of aged mice. J. Cell Biol., *26:* 779–795.

Sidman, R. L., and M. Singer 1951 Stimulation of forelimb regeneration in the newt, *Triturus viridescens*, by a sensory nerve supply isolated from the central nervous system. Am. J. Physiol., *165:* 257–260.

Singer, M. 1942 The nervous system and regeneration of the forelimb of adult *Triturus*. I. The role of the sympathetics. J. Exp. Zool., *90:* 377–400.

——— 1943 The nervous system and regeneration of the forelimb of adult *Triturus*. II. The role of the sensory supply. J. Exp. Zool., *92:* 297–316.

——— 1946 The nervous system and regeneration of the forelimb of adult *Triturus*. V. The influence of number of nerve fibers, including a quantitative study of limb innervation. J. Exp. Zool., *101:* 299–338.

Smith, K. R. 1961 The fine structure of neurons of dorsal root ganglia after stimulating or cutting the sciatic nerve. J. Comp. Neur., *116:* 103–115.

Stevens, B. J. 1964 The effect of actinomycin D on nucleolar and nuclear fine structure in the salivary gland cell of *Chironomus thummi*. J. Ultrastruct. Res., *11:* 329–353.

Warner, J. R., P. M. Knopf and A. Rich 1963 A multiple ribosomal structure in protein synthesis. Proc. Nat. Acad. Sci., *49:* 122–129.

Abbreviations

AN, accessory nucleolus
Ax, axon
Ch, channels (smooth endoplasmic reticulum)
D, droplet
DG, dense granules
DV, dense vesicles
G, Golgi apparatus
LD, lipid droplet
Lf, lipofuscin pigment body
Ly, Lysosome
M, Mitochondria
Mt, microtubules
N, nucleus
Ni, Nissl substance
Nn, nucleolonema
PA, pars amorpha
Va, vacuole

PLATE 1

EXPLANATION OF FIGURE

1 Normal sensory ganglion cell. The large cell has an eccentrically situated nucleus with nucleoli and an axon (Ax) extending from one pole. Golgi complexes (G) occur in the perinuclear region and also elsewhere in the cytoplasm. Round or oval mitochondria with a diameter of 0.2–0.5 μ are scattered throughout the cell. Short, rough-surfaced cisternae of endoplasmic reticulum occur and may be concentrated into Nissl bodies (Ni). Dense membrane-bounded structures represent lipofuscin pigment bodies (Lf). The axon hillock contains fewer organelles than other regions of the cell. Tubular channels (Ch) of smooth-surfaced endoplasmic reticulum occur in the proximal region of the axon. A few of these channels extend into the hillock and connect with the rough-surfaced cisternae bordering this area (arrows). × 14,600.

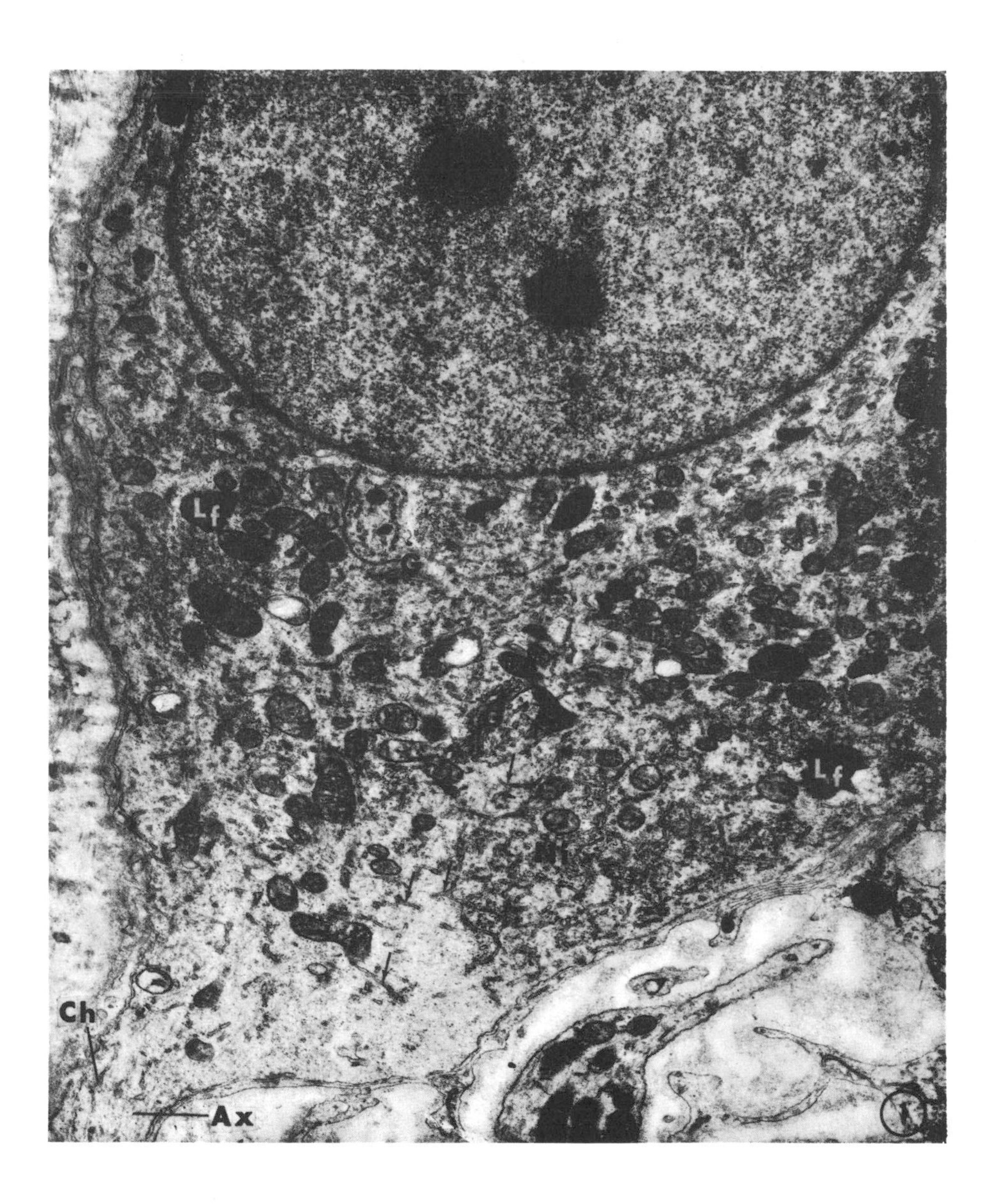
Lf
G
Lf
N
Ch
Ax

PLATE 2

EXPLANATION OF FIGURES

2 Nucleus of a normal ganglion cell illustrating a large nucleolus and smaller accessory nucleolus (AN). The nucleolus contains two round masses of fine textured material comprising the pars amorpha (PA). Fibrillar material also composes course anastomosing strands or nucleolonema (Nn). Small granules occur in the interstices and within the strands. The accessory nucleolus is composed of very finely granular and fibrillar material. The nuclear substance consists of dense granules and thin strands embedded within the electron-lucent nucleoplasm. × 23,000.

3 Normal sensory ganglion cell. The perinuclear region contains fewer organelles than elsewhere and microtubules (Mt) extend circumferentially around the nucleus in this region. Golgi complexes (G) occur external to the microtubules and may have small vesicles and dense core vesicles (DV) associated with them. A small vacuole (Va) near the Golgi apparatus contains finely granular material identical to that comprising lysosomes (Ly). Mitochondria and lipofuscin pigment bodies occur in the cytoplasm. A number of rough-surfaced cisternae of endoplasmic reticulum are concentrated to form a Nissl body (Ni). Free ribosomes are especially abundant in the cytoplasm between the cisternae. N, nucleus. × 26,000.

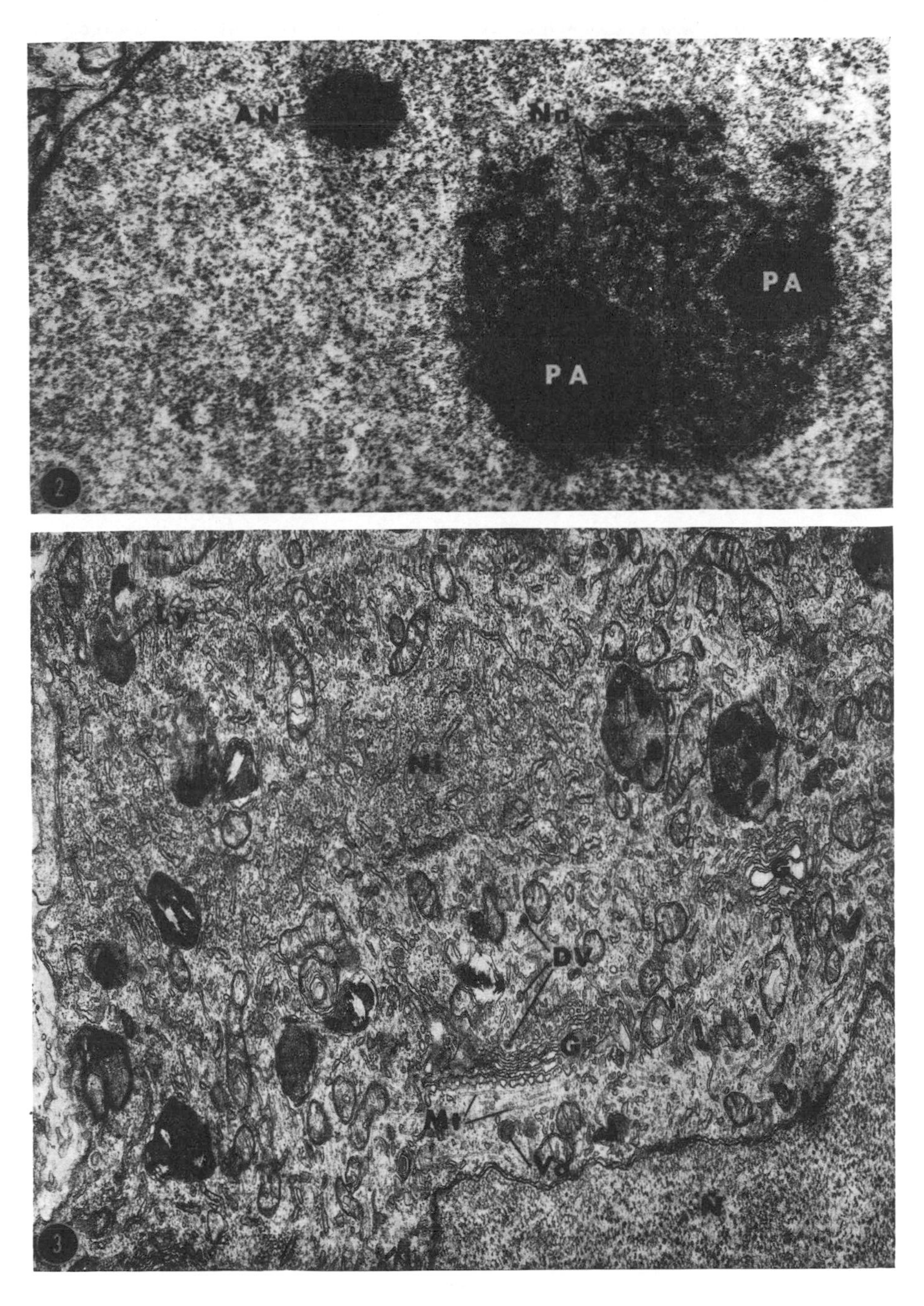
AN
Nu
PA
PA
2
Ly
Ni
DV
G
Mt
Vd
N
3

PLATE 3

EXPLANATION OF FIGURES

4 Cytoplasm of a sensory ganglion cell three days after amputation of the limb. This portion of the cell is filled with lipofuscin pigment bodies. These structures are bounded by a membrane and contain several masses of opaque material. In some, a large droplet (D) with a lucent center is present. Some contain one or more membranes (arrows) beneath the outer limiting membrane. Another type of structure containing granular material may represent a type of lysosome (Ly). Sometimes opaque material occurs within the granular matrix (structure to right of lysosome) suggesting an intermediate stage in the transformation of a lysosome to a pigment body. × 19,000.

5 Sensory ganglion cell seven days after limb amputation. Two giant mitochondria (M) with a diameter of ~ 2 μ are present in this cell. Other mitochondria are intermediate in size between the small normal sized (0.2 μ) and the large mitochondria. A Golgi apparatus (G) occurs near the nucleus (N). The cisternae of endoplasmic reticulum appear somewhat more dispersed than normal. Pigment bodies and a large lipid droplet (LD) are present. × 19,000.

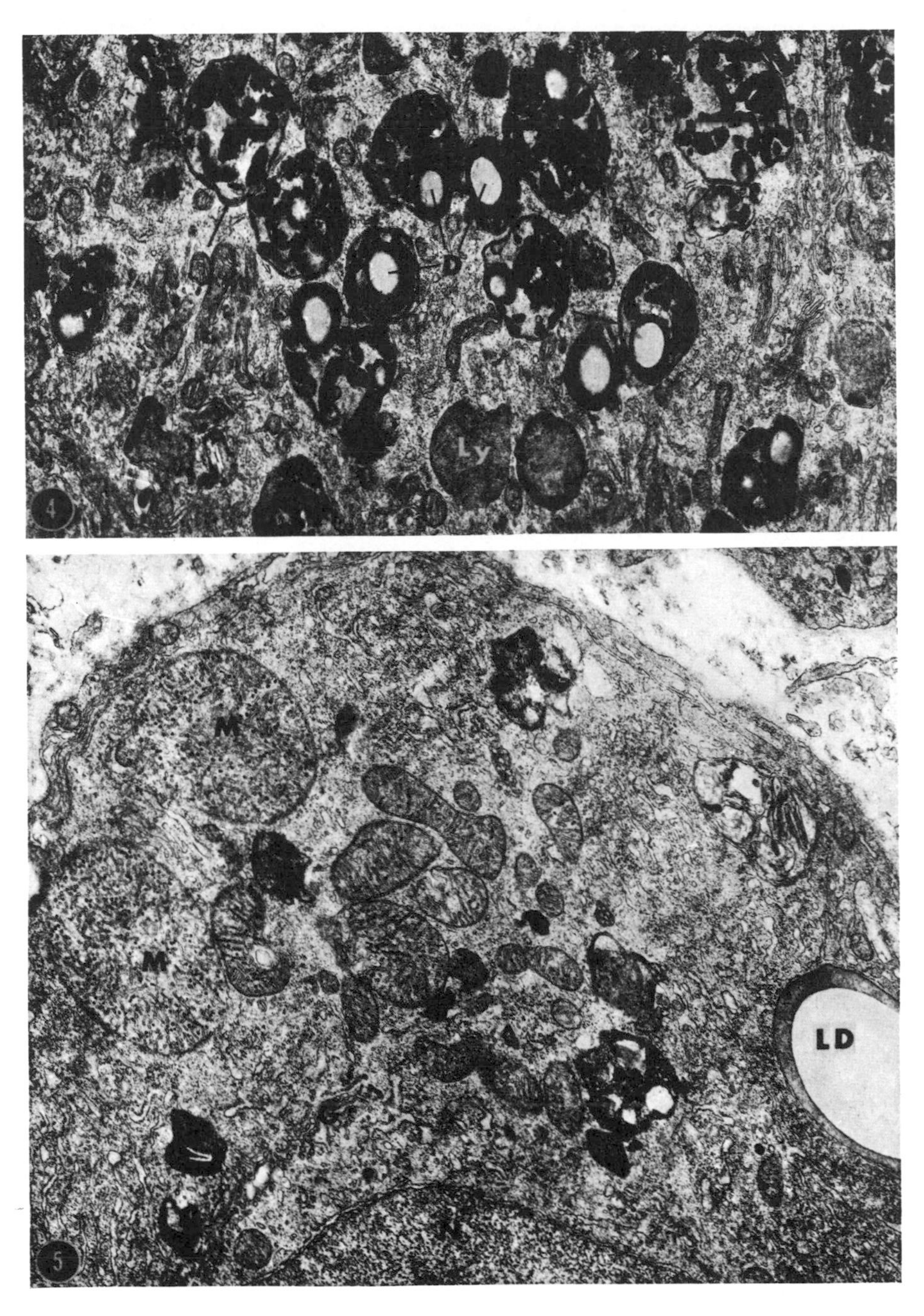
D
Ly
4
M
M
LD
5

PLATE 4

EXPLANATION OF FIGURE

6 Sensory ganglion cell three days after limb amputation. A large number of giant mitochondria occur in the cytoplasm of this cell. Comparison with normal sized mitochondria (arrows) emphasizes the great enlargement undergone by these organelles. Note the tubular cristae and opaque granules in the matrix of the mitochondria. Small Golgi complexes (G) occur in the cytoplasm between the mitochondria. The rough-surfaced endoplasmic reticulum appears diminished in amount in this cell. Ax, Axon. × 15,500.

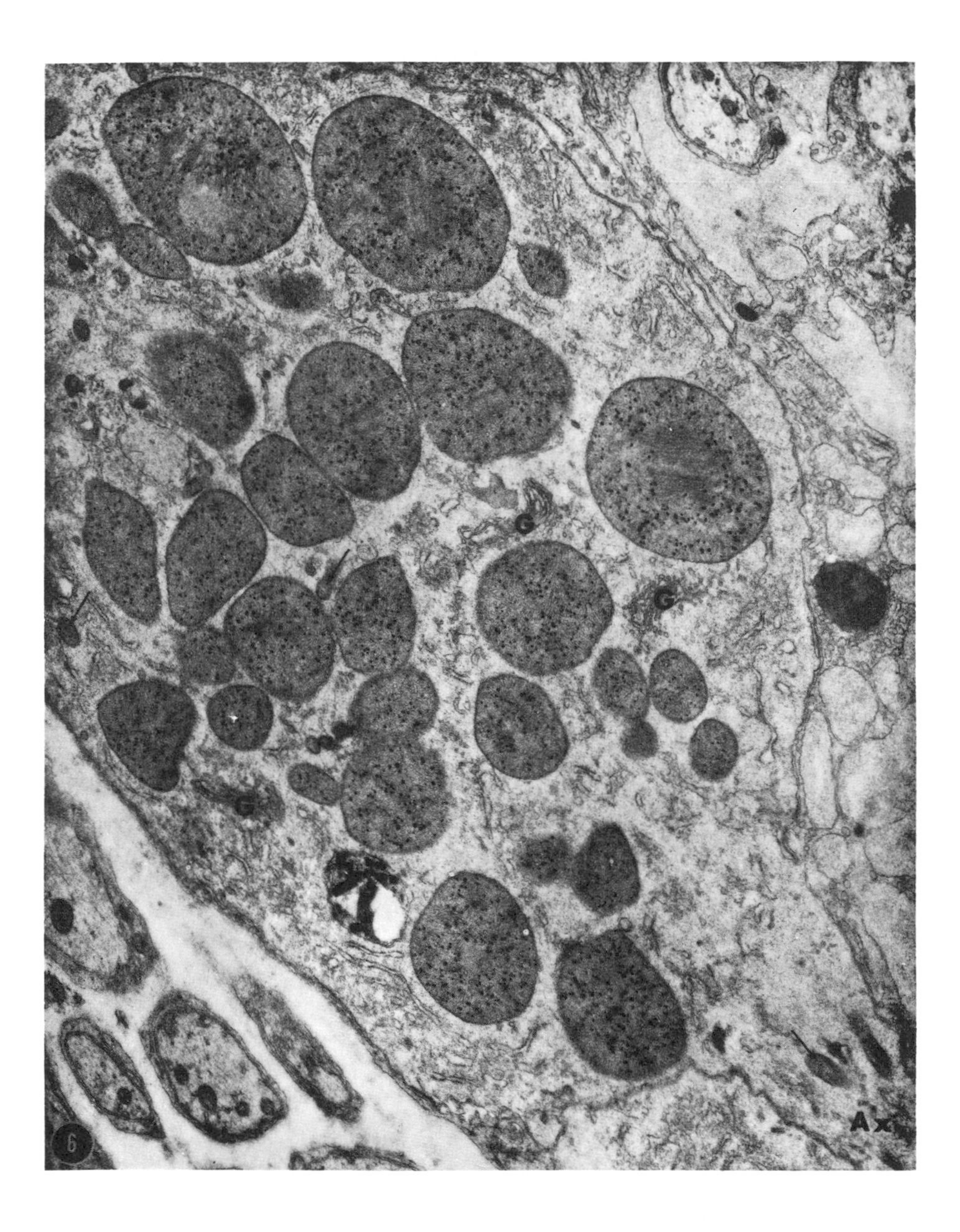
G
G
G
Ax
6

PLATE 5

EXPLANATION OF FIGURE

7 Mitochondria and Golgi complexes (G) in a sensory ganglion cell three days after limb transection. The Golgi complexes are composed of a stack of short membranous cisternae and small vesicles. Dense granules (DG), bounded by a membrane and ~ 1000 A in diameter, occur in the Golgi regions. The cristae of the large mitochondria are tubular in cross section and extend irregularly toward the center of the organelle. The central region of the lower mitochondrion is largely devoid of cristae leaving the amorphous, moderately dense matrix material. Electron-opaque granules of different sizes occur in the matrix. × 50,500.

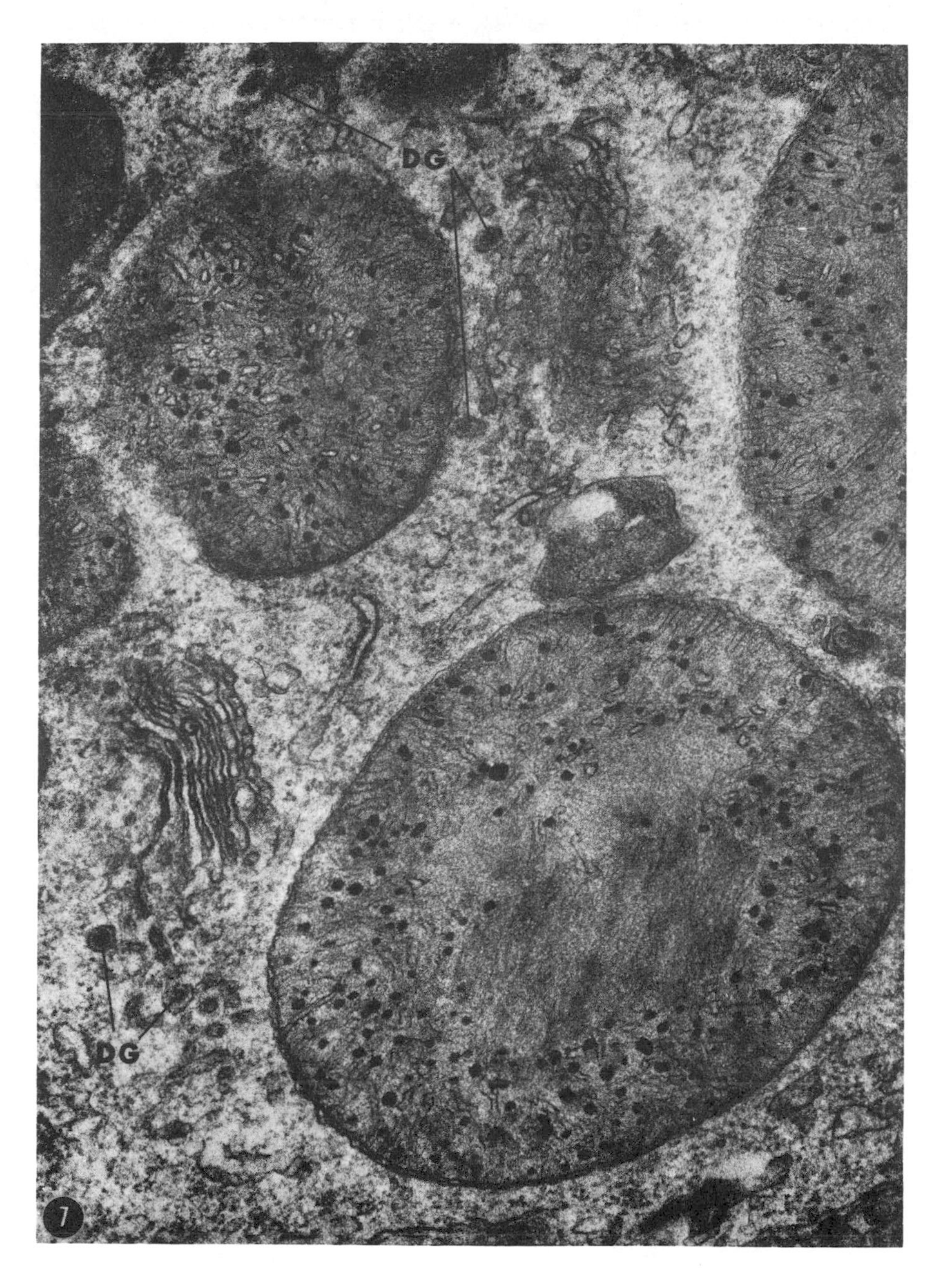
DG
G
DG
7

PLATE 6

EXPLANATION OF FIGURES

8 Ganglion cell 14 days after transection of the limb. A large portion of this cell is occupied by a cluster of mitochondria which are the same size or only slightly larger than normal. The mitochondria are tightly packed excluding most other organelles from this region. The perinuclear Golgi complexes (G) do not differ from normal. × 13,500.

9 Ganglion cell seven days after limb amputation. The cisternae of rough-surfaced endoplasmic reticulum are more dispersed than normal occupying much of the cytoplasm. The cisternae appear slightly dilated and long stretches of membrane are devoid of attached ribosomes. Free ribosomes are more abundant than normal, and may occur in clusters in the cytoplasm (arrows). N, nucleus. × 25,500.

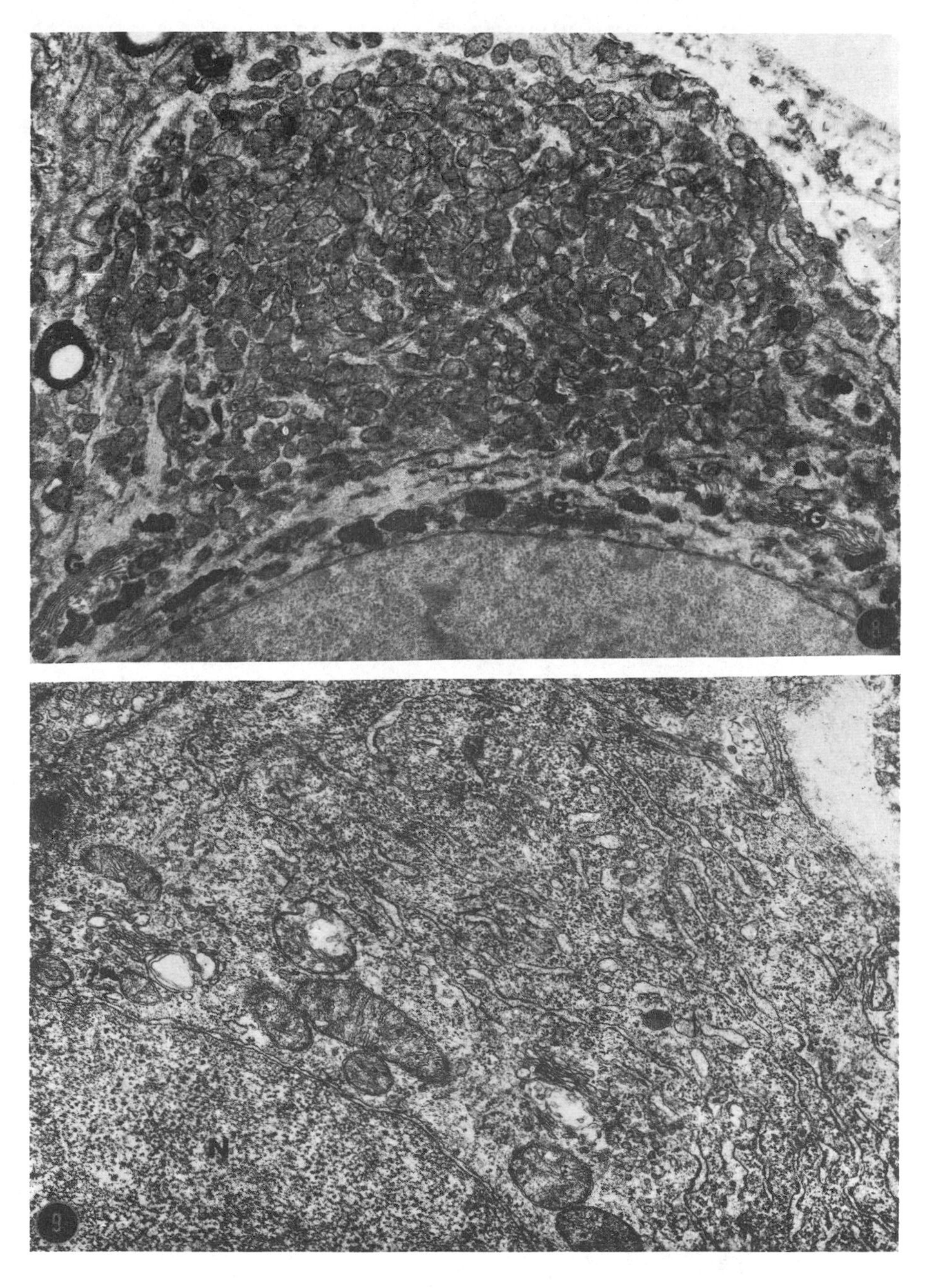
G
G
G
8
N
9

PLATE 7

EXPLANATION OF FIGURES

10 Sensory ganglion cell showing marked chromatolysis 21 days after limb transection. Cisternae of endoplasmic reticulum are reduced in number and confined to small peripheral clumps. The cytoplasm formerly occupied by endoplasmic reticulum contains amorphous material of low density within which is embedded dense granules. Mitochondria do not seem as common although Golgi complexes (G) persist. The contour of the nucleus bears undulations and the nuclear contents occur both dispersed and within small clumps. × 17,500.

11 Axon (Ax) and hillock region of a ganglion cell 14 days after limb amputation. The axon hillock contains more organelles than usual including mitochondria and rough-surfaced endoplasmic reticulum. Mitochondria extend into the proximal portion of the axon. The axon contains tubular channels (Ch) of smooth endoplasmic reticulum which contain moderately dense material and seem more numerous than normal. Microtubules may also be discerned in the axon. × 27,000.

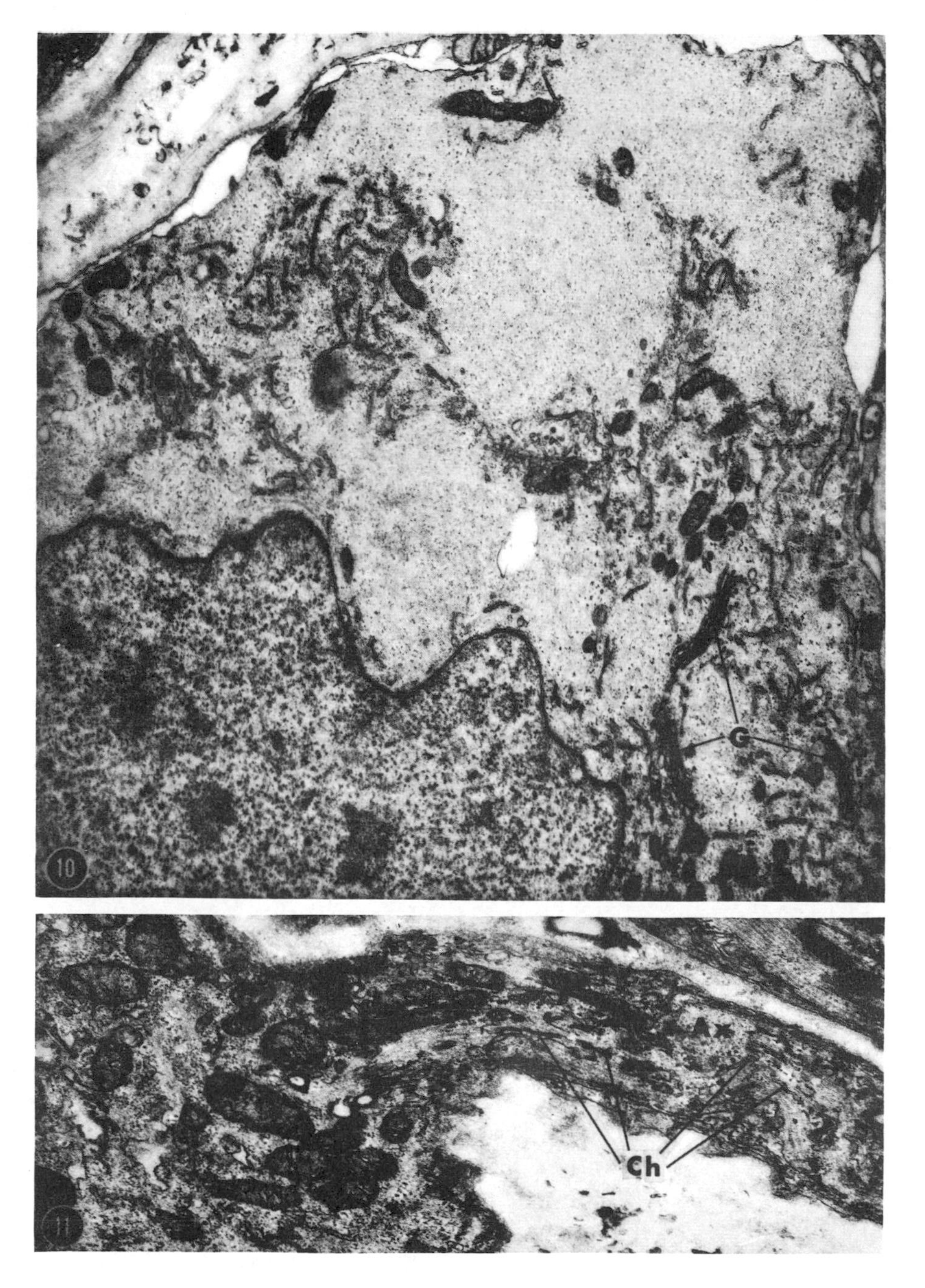
10
G
11
Ch

Influence on Amphibian Limb Regeneration of Hormones and Other Agents

Hormonal and Nutritional Requirements for Limb Regeneration and Survival of Adult Newts

ROY A. TASSAVA

The adult newt, *Notophthalmus (Triturus) viridescens,* requires pituitary hormones for survival and normal limb regeneration (Richardson, '45; Dent, '67; Connelly et al., '68). The exact nature of the essential pituitary hormones and their mechanism of action on survival and limb regeneration are unknown.

Schotté ('61) has suggested that in typical limb regeneration the newt pituitary is stimulated to produce ACTH (adrenocorticotrophic hormone) by the stress of amputation; the ACTH then activates the adrenal gland which produces steroid hormones. The adrenal hormones are thought to be essential for proper wound healing during the first six days after limb amputation and this is prerequisite for initiating regeneration (Schotté and Hall, '52). However, evidence in support of an ACTH-adrenal mechanism in newt limb regeneration and survival is not compelling. Thus, ACTH injections into hypophysectomized adult newts brought about limb regeneration and enhanced survival (Schotté and Chamberlain, '55), but adrenal steroids did not enhance survival or limb regeneration of hypophysectomized adult newts (Schotté and Bierman, '56). Schotté and Chamberlain ('55) found that ACTH injections into intact newts caused a temporary delay in limb regeneration, yet Bragdon and Dent ('54) observed normal limb regeneration in intact newts injected with ACTH or cortisone.

There is other evidence which makes doubtful the possibility of an ACTH involvement in limb regeneration. A pituitary-adrenal axis apparently exists in the bullfrog (*Rana catesbiana*) (Piper and DeRoos, '67), nevertheless the adrenal of both anuran (Hanke and Weber, '65) and urodele amphibians (Wurster and Miller, '60) displays considerable independence from the pituitary gland. A hypophysectomized adult newt with an ectopic pituitary gland will regenerate amputated limbs in a normal fashion (Schotté and Tallon, '60), yet in other vertebrates, experiments have shown that the ectopically transplanted pituitary (removed from its hypothalamic connections) does not respond well to stress (Van Dongen et al., '66; Mangili et al., '66; Purves and Sirett, '67). Also, it is known in mammals that the response to stress is nearly immediate, occurring within seconds of the onset of stimulation and, furthermore, that denervation of a limb or ablation of the spinal cord prevents the ACTH response to stimuli applied to the denervated territory (Fortier, '66). When the lumbo-sacral spinal cord of adult *Ambystoma* is ablated, amputated hind limbs and tail nevertheless regenerate (Liversage, '59). Furthermore, when limb and spinal cord segments are transplanted to the dorsal fin, followed by spinal cord ablation of the host and later amputation of the limb, normal regeneration follows. In these ablation experiments, neural stimulation of endocrine activities (especially the pituitary-adrenal axis) was prevented (Liversage, '59).

Hormones other than ACTH have been investigated in newt limb regeneration. Connelly et al. ('68) found that prolactin, when combined with thyroxine, was very effective in enhancing survival and promoting limb regeneration in hypophysectomized adult newts. Prolactin alone, in larger quantities, was less effective and thyroxine alone was completely ineffective. Richardson ('45) found that Antuitrin G, a crude growth hormone preparation, supported limb regeneration in hypophysectomized newts but not as well as Antuitrin G combined with thyroxine. Wilkerson ('63) obtained excellent limb regeneration in hypophysectomized adult newts with growth hormone (NIH) even when injections were begun 14 days after amputation and hypophysectomy. However, the growth hormone (GH) used by Wilkerson ('63) contained prolactin as a contaminant in an amount equivalent to that used by Berman et al. ('64) which stimulated growth and inhibited metamorphosis in the frog tadpole. The prolactin contamination in the GH was also comparable to the amount of prolactin which will induce the red eft, the land stage of the newt, to migrate to water (Grant and Grant, '58). That prolactin may play a role in newt limb regeneration was further suggested by Niwelinski ('58) and Waterman ('65) who found an enhancement of limb regeneration by prolactin injections into intact newts, and by the report of Chadwick and Jackson ('48) that prolactin increases mitotic activity in newt epidermis.

It has been suggested that pituitary hormones may act during the growth phase of limb regeneration (Hay, '56, '66) instead of only during the wound healing phase (Schotté and Hall, '52). Hay ('56, '66) points out that the best regenerative response in hypophysectomized newts occurred when frog adrenals were implanted *15 days* after amputation (Schotté and Lindberg, '54) and when growth hormone injections were begun 14 days after amputation (Wilkerson, '63). The data of Schotté and Hall ('52) also indicate that pituitary hormones may influence the growth of the blastema. Thus, when hypophysectomy of adult newts was delayed until 14 days after limb amputation, all of the limbs regenerated but in 79% of the cases regeneration was abortive or delayed (Schotté and Hall, '52). At 14 days after amputation, wound healing is essentially complete, as is dedifferentiation (Hall and Schotté, '51), yet hypophysectomy apparently has an adverse effect on the 14 day regenerate (Schotté and Hall, '52).

It is not clear, therefore, which pituitary hormones are essential to newt survival and limb regeneration; whether these pituitary hormones act through the adrenal gland; which stage(s) of limb regeneration is influenced by pituitary hormones; and finally, whether pituitary hormones are absolutely essential to the initiation of limb regeneration. Therefore, experiments described in this paper were designed to determine: (1) whether limb regeneration absolutely requires pituitary hormones, particularly when newts are in good nutritional condition at the time of hypophysectomy; (2) whether the 14 day early bud regenerate of an adult newt will continue normal growth and differentiation in the absence of pituitary hormone; and finally (3) which hormones or combination of hormones are most effective in prolonging survival and promoting limb regeneration of hypophysectomized adult newts.

GENERAL METHODS

Adult newts, *Notophthalmus* (*Triturus*) *viridescens*, were obtained from Lewis Babbitt, Petersham, Massachusetts. Except for some newts in Series I (Nutritional Effects), all newts were fasted for two weeks prior to hypophysectomy and were not fed during the experimental period. Using fine pointed watchmaker's forceps, a small flap of the sphenoid bone was loosened directly ventral to the pituitary. The anterior end of the bone flap was then lifted away while the entire pituitary gland was removed by gentle suction with a fine glass pipette (Connelly et al., '68). The flap of bone was then replaced. If any hypophysectomy was doubtful due to injury of the gland or to excess bleeding, that particular newt was immediately discarded.

Newts which are completely hypophysectomized do not shed the outer layer of their epidermis because of a thyroxine deficiency resulting from the absence of thyroid stimulating hormone (TSH) (Dent, '66). These newts develop a rough, black skin, quite different from the dark, smooth and slippery skin of newts subjected to an excess melanophore stimulating hormone (MSH) secretion. Any hypophysectomized newts which did not exhibit complete lack of molting were discarded.

Heads and limbs were fixed in Bouin's fluid, decalcified, dehydrated, cleared in methyl salicylate, embedded in paraffin, and serially sectioned at 10 μ. Sections of heads were stained by the PAS technique or Herlant's tetrachrome (Pearse, '60). Sections of limbs were stained with hematoxylin-eosin, iron-hematoxylin, or Masson's trichrome (Merchant et al., '63).

EXPERIMENTAL

Series I. Nutritional effects

This series was designed to compare the ability to regenerate limbs and the length of survival of: (1) newts which were in good nutritional condition at the time of complete hypophysectomy; (2) newts which were fasted for two weeks prior to *complete* hypophysectomy; and (3) newts which were fasted for two weeks prior to *incomplete* hypophysectomy.

A. *Methods.* Forty-four adult newts, weighing from 1–2 gm, were randomly selected from planted aquaria and maintained in 2-quart finger bowls at a density of four newts per bowl. All newts in this and the following series were maintained at 22°C ± 2°C. Of the 44 newts, 16 were randomly selected and fed beef liver maximally each day for two weeks. The remaining 28 newts were fasted for the same two week interval. After the two week interval of feeding or fasting, the individual body weights of the newts fed for two weeks were significantly greater than the individual body weights of the newts which were fasted for the same two week period (Mann-Whitney U Test, 0.01 level of significance). On the average, each fed newt weighed 0.32 gm more than each fasted newt. All 16 of the fed newts and 16 of the fasted newts were then completely hypophysectomized. The remaining 12 fasted newts were partially hypophysectomized. Using the suction technique of hypophysectomy, it is much easier to remove the entire pituitary than to leave a fragment. Therefore, for these partial hypophysectomies, only forceps were used. Care was taken to cut the pituitary gland *in situ* and to remove all but one or two fragments of the anterior lobe.

Both forelimbs of the newts in all three groups were amputated just distal to the elbow at five days post-operation (complete or partial hypophysectomy). None of these newts were fed during the course of the experiment. The day of death of each newt was recorded. Limbs and heads were fixed either immediately or within a few hours after death or were randomly sampled from the surviving newts.

Of the 16 newts which were fed for two weeks prior to hypophysectomy, eight were still surviving 26 days post-hypophysectomy (21 days post-amputation). Both forelimbs and the head of each of these eight newts were fixed at that time (26 days post-hypophysectomy) and examined histologically.

B. *Results.* Of the newts in Series I which were fed maximally for two weeks prior to hypophysectomy, 12 of 16 were alive 20 days post-hypophysectomy. One of the 16 newts which were fasted for two weeks prior to hypophysectomy was alive 20 days post-hypophysectomy. Two of the 12 fasted newts which were incompletely hypophysectomized were alive 20 days post-hypophysectomy. Figure 1 compares the per cent survival of these three groups. The 12 surviving, maximally-fed newts represent a significantly greater number than the one surviving, fasted, hypophy-

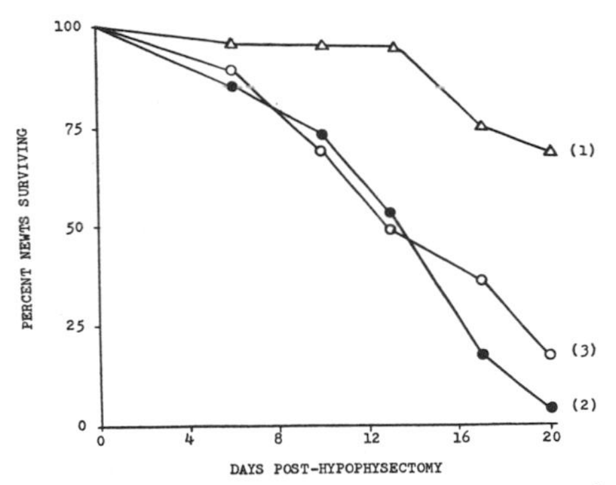

Fig. 1 A comparison of the per cent survival of (1) hypophysectomized newts fed for two weeks prior to hypophysectomy, (2) hypophysectomized newts fasted for two weeks prior to hypophysectomy, and (3) partially hypophysectomized newts fasted two weeks prior to patrial hypophysectomy.

sectomized newt or the two surviving, fasted, partially hypophysectomized newts (X^2 Test, 0.01 level of significance). Of the 16 hypophysectomized newts which were fed for two weeks prior to hypophysectomy, eight survived to day 26 post-hypophysectomy and were sacrificed. The 16 amputated limbs of these eight newts (day 21 post-amputation) revealed regeneration blastemas in every case (fig. 8; see table 1). Examination of serial sections of the heads of these eight surviving newts revealed no pituitary fragments (compare figs. 10, 11). In contrast, 10 of 12 of the heads of the partially hypophysectomized newts contained small pituitary fragments (fig. 12), yet, a significantly smaller number of these newts survived than did newts which were completely hypophysectomized and had been previously well fed. These ten newts with small pituitary fragments nevertheless failed to molt. Thus, the small pituitary fragment may not have contained the essential cell type(s) for newt survival. Histological examination of four limbs of the fasted completely hypophysectomized newts and six limbs of the fasted incompletely hypophysectomized newts which were fixed 20–26 days post-hypophysectomy (15–21 days regeneration) revealed regeneration blastemas although retarded as compared to normal regenerates (table 1; fig. 9).

In summary, 16 of 16 limbs of maximally fed newts which were amputated five days post-hypophysectomy exhibited clear indications of early stages of regeneration when examined histologically 26 days post-hypophysectomy. Thus, limbs of newts which are amputated after hypophysectomy do initiate regeneration in the absence of the pituitary gland. Furthermore, survival depends upon the nutritional state of the newt at the time of hypophysectomy.

Series II. Pituitary influence on growth of the 14 day regenerate

The results of Series I clearly demonstrate that limb regeneration can proceed at least to the medium bud stage in the complete absence of pituitary hormones. This series was designed to determine whether regenerates which have reached the bud stage in the presence of pituitary

TABLE 1

Delay of regeneration produced by hypophysectomy

Days after amputation [1]	Stages of regeneration attained		
	Intact newts [2]	Hypophysectomized previously fed newts	Hypophysectomized previously fasted newts [3]
10–13	Accumulation blastema		
14–15	Early bud	Accumulation blastema	
16–20	Medium bud	Early bud	Accumulation blastema
21–25	Late bud and cartilage differentiation	Medium bud	Early bud
26–29	Palette		Medium bud
30 +	Digit		

[1] Limbs were amputated five days post-hypophysectomy.
[2] Limb regeneration of intact newts injected with ACTH (UpJohn) corresponds to that of intact untreated newts. Regeneration stages after Singer ('52).
[3] Limb regeneration of hypophysectomized newts treated with saline, prolactin alone, ACTH, and/or thyroxine corresponds to that of hypophysectomized, previously fasted newts.

hormones can continue to grow and differentiate in a normal fashion after removal of the pituitary hormones.

A. *Methods.* The right or the left forelimb (chosen randomly) of each of 36 newts, previously fasted for two weeks, was amputated through the distal portion of the humerus. At 16 days post-amputation these newts were randomly divided into two groups. The limbs of two newts were sampled at this time for histology to ascertain the status of regeneration. Both limbs exhibited early bud regenerates. The newts in one group ($n = 20$) were hypophysectomized while the remaining 16 newts were sham-operated. The sham-operation consisted of removing a portion of the bone over the sella turcica without removing the pituitary. The bone was then replaced. At eight days post-hypophysectomy or post-sham (22 days post-amputation), six of the newts from each group were randomly selected, sacrificed and the amputated limb and the head of each newt were fixed and prepared for histological examination. Limbs were oriented randomly in paraffin, serially sectioned longitudinally at 10 μ, and stained with hematoxylin and eosin. The area and the length of the regeneration blastema were determined for each of five sections of each limb. The image of the section to be measured was projected on to graph paper from a constant height. The image of the blastema of each section sampled was then traced on to the graph paper with a fine pointed pencil. The number of squares representing the blastema was counted for each section sampled (5 sections for each limb). The five sections to be sampled were determined by selecting the approximately largest section (usually in the center of the limb), and measuring the area of the blastema of that section. Two sections, 100 μ apart, on both sides of this largest section were also sampled. Thus the five sections sampled, each 10 μ thick and spaced at 100 μ intervals, covered a total distance through the blastema of 450 μ. The mean blastema area and length for each limb were determined (mean of 5 sections) and the means were compared statistically for the two groups using the non-parametric Mann-Whitney U Test.

In addition to the six newts fixed eight days after hypophysectomy, seven of the 20 hypophysectomized newts died during the course of the experiment. The seven surviving hypophysectomized newts were sacrificed 16 days after hypophysectomy (30 days post-amputation) and their limbs and heads prepared for histology. Area measurements were made of these blastemas as described above. The limbs of the eight surviving sham-operated newts were also fixed at this time and prepared for histology. Area measurements were also made of these blastemas. These mean blastema areas, 16 days post-hypophysectomy or post-sham operation, were compared statistically.

B. *Results.* The results of Series II are summarized in figure 2, figure 3, and table 2. The mean blastema areas of the sham-operated newts were significantly greater (0.01 level) than the mean blastema areas of the hypophysectomized newts at eight days post-hypophysectomy (15.90 squares vs. 9.25 squares) and also at 16 days post-hypophysectomy (22.47 squares vs. 15.22 squares). At eight days post-operation, the blastemas of the sham-operated newts were also significantly longer (0.01 level) than the blastemas of the hypophysectomized newts (17.5 units vs. 11.5 units). Histological and gross observations also revealed that the regenerates of the sham-operated newts were in more advanced stages of differentiation. Digit differentiation was apparent in five of the six regenerates of the sham-operated newts at 16 days post-sham-operation whereas only one of the six regenerates of the hypophysectomized newts exhibited digit differentiation 16 days post-hypophysectomy. Cells in mitosis could be observed in the regenerates of the hypophysectomized newts at eight days post-hypophysectomy; therefore growth was not completely stopped. Furthermore, the mean blastema areas were greater at 16 days than at eight days post-hypophysectomy suggesting that some growth did occur (table 2; fig. 3) in the complete absence of the pituitary.

In summary, the results of Series II demonstrate that the normal growth of the blastema requires pituitary hormones. Hypophysectomy has an adverse effect on limb regeneration even after the wound

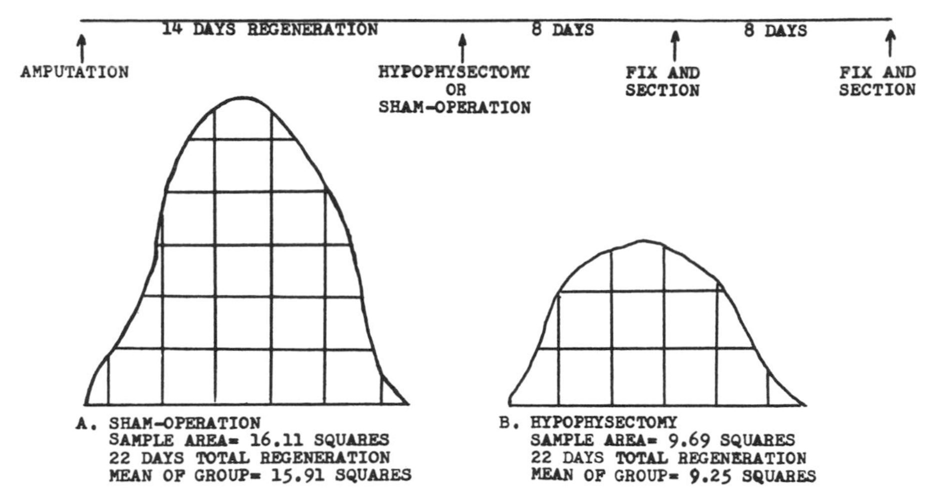

Fig. 2 The influence of hypophysectomy on newt limb blastema area. Fasted newts were hypophysectomized or sham-operated 14 days after limb amputation. At eight days and also at 16 days after hypophysectomy or sham-operation, the limbs were sampled for sectioning and staining. In A and B the section areas were obtained by projecting the blastema image onto graph paper from a constant height and tracing the blastema outline. The mean blastema areas of the sham-operated newts were significantly larger (0.01 level of significance; Mann-Whitney U Test).

healing and dedifferentiation phases are complete.

Series III. Hormone replacement therapy

The results of Series I and Series II demonstrate that pituitary hormones are more important to the later growth phase of limb regeneration than to the early, wound healing phase. In addition factors which promote survival of hypophysectomized newts also promote limb regeneration (i.e. good nutritional condition). These facts do not lend support to an ACTH-adrenal mechanism but suggest instead that pituitary hormones which promote survival also make it possible for normal limb regeneration to proceed. Therefore, Series III was designed to determine which hormones or combination of hormones effectively enhanced survival and/or limb regeneration.

A. *Methods.* Fasted newts were hypophysectomized and randomly divided into groups. Different groups (tables 3, 4) of hypophysectomized newts were either not treated or received ectopic pituitary glands from donor newts or donor axolotls, *Ambystoma mexicanum,* or intraperitoneal injections of either water, saline, prolactin (NIH), growth hormone (NIH-GH) or adreno-corticotropic hormone (UpJohn-ACTH). In addition some hypophysectomized newts were treated with thyroxine, with and without prolactin or ACTH. The thyroxine was dissolved in the water in which the newts were maintained at a concentration by weight of one part thyroxine/10 million parts of water (1×10^{-7} conc. in water). The newts were changed to fresh solution every other day. The prolactin and GH were dissolved in 0.9% saline and the ACTH in sterile water at concentrations such that 0.1 cm^3 of solution contained the desired amount of hormone. For the pituitary transplants, the host newt was anesthetized in MS 222 and after hypophysectomy of the host, the donor (either newt or axolotl) was decapitated and its entire pituitary (neural, intermediate and anterior lobes) was quickly removed and transplanted to the musculature of the lower jaw of the host.

Plastic disposable syringes with 27 gauge hypodermic needles were used for

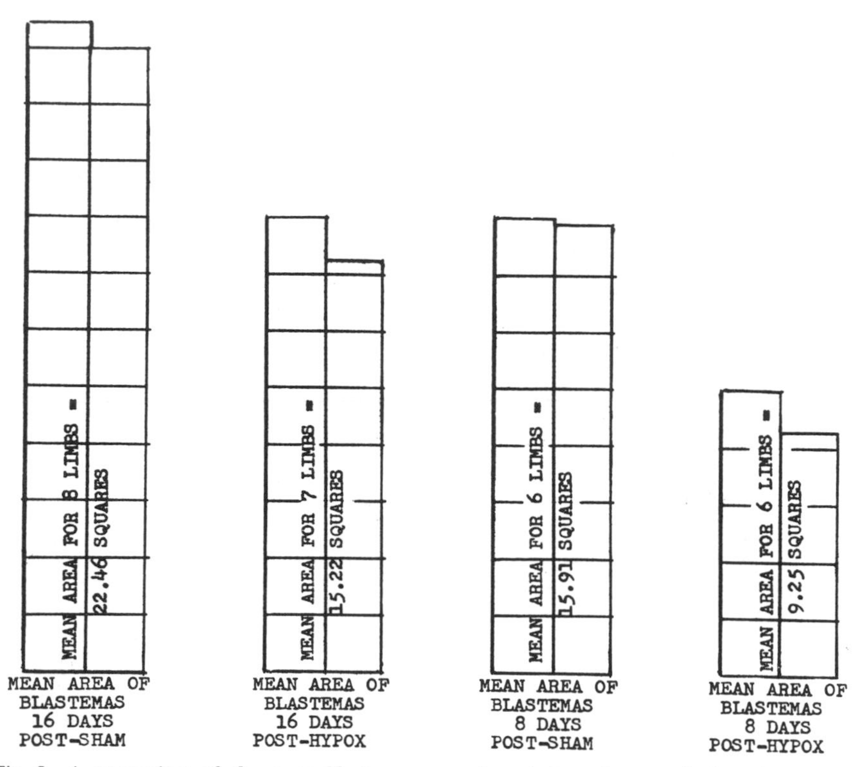

Fig. 3 A comparison of the mean blastema areas of newts hypophysectomized or sham-operated at 14 days after limb amputation and fixed eight days post-operation or at 16 days post-operation.

TABLE 2

The effect of hypophysectomy on blastema area

Mean area of blastema sections 8 days post-operation (22 days regeneration)				Mean area of blastema sections 16 days post-operation (30 days regeneration)			
Hypophysectomy		Sham [1]		Hypophysectomy		Sham	
Limb	Mean area of 5 sections	Limb	Mean area of 5 sections	Limb	Mean area of 5 sections	Limb	Mean area of 5 sections
1	1.86	1	11.72	1	11.26	1	16.47
2	5.07	2	13.18	2	11.87	2	17.20
3	9.55	3	16.13	3	14.19	3	17.78
4	10.37	4	16.30	4	14.49	4	19.95
5	11.71	5	17.11	5	16.21	5	25.46
6	16.93	6	21.03	6	18.21	6	25.51
				7	20.26	7	27.63
						8	29.68
Mean = 9.25		Mean = 15.91		Mean = 15.22		Mean = 22.46	

[1] The mean blastema areas (5 sections/limb) of the individual limbs of the sham-operated newts are significantly greater than the mean blastema areas (5 sections/limb) of the individual limbs of the hypophysectomized newts, both at eight days post-operation and at 16 days post-operation (Mann-Whitney U Test — 0.01 level of significance).

injections. Injections into hypophysectomized newts were begun either five days after hypophysectomy, at which time both forelimbs were amputated, or at ten days after hypophysectomy (delayed injections) and continued every two days until the end of the experiment.

Limbs and heads of newts from all groups were sampled for histological examination at various times during the experiment. The number of surviving newts of each group was recorded and compared statistically using the X^2 Test on days 20 and 24 post-hypophysectomy.

The hormones and control solutions with which hypophysectomized newts were treated are listed in tables 3 and 4 along with the number of newts treated, and the effect of the treatment on limb regeneration and survival of hypophysectomized newts. The amounts and activities of the contaminating hormones present in the hormone preparations used are listed in table 5. It should be noted that two growth hormone dosages were used, 0.3 mg and 0.03 mg GH/newt/2 days. The growth hormone (NIH) used in these experiments contained ten times more prolactin in each mg of preparation than did the growth hormone (NIH) used by Wilkerson ('63). Thus, the smaller amount of growth hormone (0.03 mg) still contained an amount of prolactin (0.016 U) equivalent to that given with the larger amount of GH (0.3 mg) by Wilkerson ('63).

Hypophysectomized euryhaline fish will survive when maintained in dilute saline and, in these fish, prolactin plays a role in water balance (Ball and Ensor, '67). Therefore, to test whether deaths of hypophysectomized newts were due to loss of sodium, control, untreated hypophysectomized newts, and hypophysectomized newts treated with prolactin, were provided NaCl in the aquarium water (table 3).

ACTH (1U/newt every other day) dissolved in water, was also administered to eight *intact* newts for a period of 24 days. Water (0.1 cm^3/newt) was administered to another eight intact newts. Four days after beginning the injections, both forelimbs of all 16 newts were amputated through the elbow. The status of limb regeneration of these newts was observed grossly until day 30 post-amputation.

B. *Results.* The results of Series III are summarized in tables 3 and 4 and figures 4, 5, 6, and 7. Untreated or saline-treated hypophysectomized adult newts begin to die approximately one week after hypophysectomy and few survive beyond 24 days post-hypophysectomy. A similar survival pattern is observed when hypophysectomized newts are treated with (1) prolactin alone (0.015 U/newt) (fig. 4), (2) prolactin + NaCl (fig. 5), (3) thyroxine alone (fig. 6), (4) ACTH (fig. 7), and (5) ACTH + thyroxine (tables 3, 4). The thyroxine and ACTH were no more effective when treatment was delayed until ten days post-hypophysectomy (table 4).

Hypophysectomized adult newts treated with ACTH do not molt and their skin becomes rough and turns uniformly dark. In addition, one observes the typical melanophore response of ACTH (Geschwind, '67) so that these ACTH treated newts become extremely black by two weeks post-hypophysectomy. The skin of hypophysectomized newts treated with thyroxine is smooth and has a normal olive-green appearance. Hypophysectomized newts treated with saline, ACTH (UpJohn) and/or thyroxine are sluggish and will not eat, even when repeatedly offered fresh beef liver. It should be recalled that none of the newts in these investigations were fed during the course of the experiments.

The treatments effective in enhancing survival (0.01 level of significance, X^2 Test) and limb regeneration of hypophysectomized newts were: (1) prolactin + thyroxine, (2) growth hormone, (3) ectopic pituitary grafts of newts and (4) ectopic pituitary grafts of larval axolotls (*Ambystoma mexicanum*). A prolactin-thyroxine combination significantly enhanced survival of hypophysectomized newts even when treatment was delayed until ten days post-hypophysectomy. Both quantities of growth hormone given (table 3) were effective and it should be noted that the smaller quantity (0.03 mg/newt) nevertheless contained a significant prolactin contamination (table 5). The growth hormone also contained more TSH as a contaminant than did the prolactin (table 5). The lower thyroxine concentration (1×10^{-8}), when

TABLE 3

The effect of hormone treatments on limb regeneration and survival of hypophysectomized newts when treatment is begun five days post-hypophysectomy

Treatment	No. of newts	Limb regeneration [1]	Survival [2]
Prolactin (0.015 U/newt/2 days) and Thyroxine (1×10^{-7} conc. in water) [3]	23	Normal	+
Growth Hormone (0.03 mg/newt/2 days)	15	Normal	+
Growth Hormone (0.3 mg/newt/2 days)	6	Normal	+
Ectopic Newt Pituitary (1 pit/newt)	10	Normal	+
Ectopic Axolotl Pituitary (2 pit/newt)	16	Normal	+
Prolactin (0.015 U/newt/2 days)	20	Delayed	—
Thyroxine (1×10^{-7} conc. in water)	17	Delayed	—
ACTH (1 U/newt/2 days)	15	Delayed	—
ACTH (1 U/newt/2 days) and Thyroxine (1×10^{-7} conc. in water)	12	Delayed	—
Saline (0.9% 0.1 cm^3/newt/2 days)	30	Delayed	—
Prolactin (0.015 U/newt/2 days) and NaCl (3.6 g/liter of aquarium water)	9	Delayed	—
Prolactin (0.015 U/newt/2 days) and NaCl (0.36 g/liter of aquarium water)	8	Delayed	—
NaCl (3.6 g/liter of aquarium water)	6	Delayed	—

[1] Normal regeneration refers to the amount of blastema cells comparable to that found in a regenerate of a normal (intact) newt at the same amputation age. Delayed regeneration refers to amputated limbs which may exhibit some blastema cells but an amount not comparable to that found in a regenerate of a normal (intact) newt at that same amputation age. Complete absence of regeneration exemplified by dermal pad formation and normal skin differentiation (skin gland formation) over the amputation surface was not observed. Limbs were amputated five days post-hypophysectomy.

[2] Survival classed as positive (+) refers to *over* 80% survival at 21 days post-hypophysectomy. Survival classed as negative (—) refers to *less* than 20% survival at 21 days post-hypophysectomy.

[3] Another eight hypophysectomized newts received prolactin (0.015 U/newt/2 days) + thyroxine (1×10^{-8} conc. in water). This experiment is depicted in figure 5.

TABLE 4

The effect of hormone treatment on limb regeneration and survival of hypophysectomized newts when treatment is begun ten days post-hypophysectomy [1]

Treatment	No. of newts	Limb regeneration	Survival
Prolactin (0.015 U/newt every other day) and Thyroxine (1×10^{-7} conc. in water)	22	Normal	+
Thyroxine (1×10^{-7} conc. in water)	9	Delayed	—
ACTH (1.0 U/newt every other day)	12	Delayed	—
0.9% Saline (0.1 cm^3/newt every other day)	9	Delayed	—
No treatment	10	Delayed	—

[1] Limbs were amputated five days post-hypophysectomy.

combined with prolactin, appeared to be equivalent to the higher concentration (1×10^{-7}) in promoting survival and limb regeneration of hypophysectomized newts (fig. 5).

It should be noted that figures 1, 4, 5, 6, and 7 each represents a separate experiment and statistical comparisons were made only within each experiment. This is important because the newts were received in several shipments. Thus, the nutritional state of the newts receiving the experimental treatments probably varied from group to group, even though the newts were

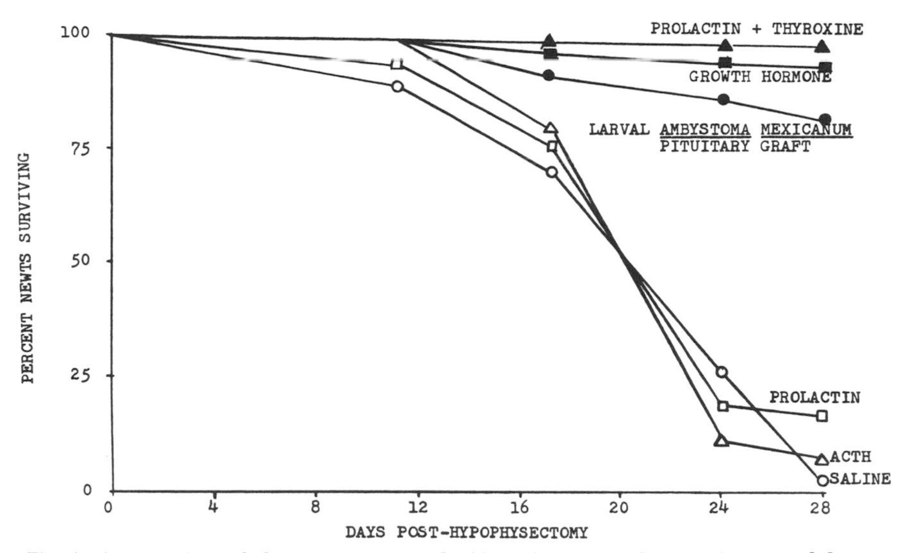

Fig. 4 A comparison of the per cent survival of hypophysectomized newts given growth hormone (0.03 mg/newt/2 days), prolactin (0.015 U/newt/2 days) + thyroxine (1×10^{-7} conc.), prolactin alone (0.015 U/newt/2 days), ACTH (1 U/newt/2 days), 0.9% saline (0.1 cm^3/newt/2 days) and ectopic, *Ambystoma mexicanum* pituitary grafts (2 pituitaries/newt).

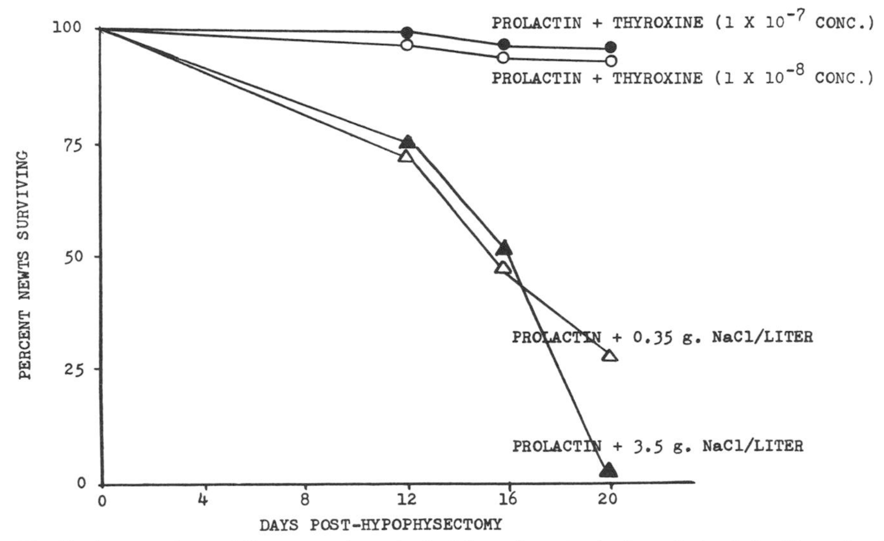

Fig. 5 A comparison of the per cent survival of hypophysectomized newts treated with prolactin (0.015 U/newt/2 days) + thyroxine (1×10^{-7} conc.), prolactin (0.015 U/newt/2 days) + thyroxine (1×10^{-8} conc.), prolactin (0.015 U/newt/2 days) and maintenance in aerated water containing 0.35 gm NaCl/liter, and prolactin (0.015 U/newt/2 days and maintenance in aerated water containing 3.5 gm NaCl/liter.

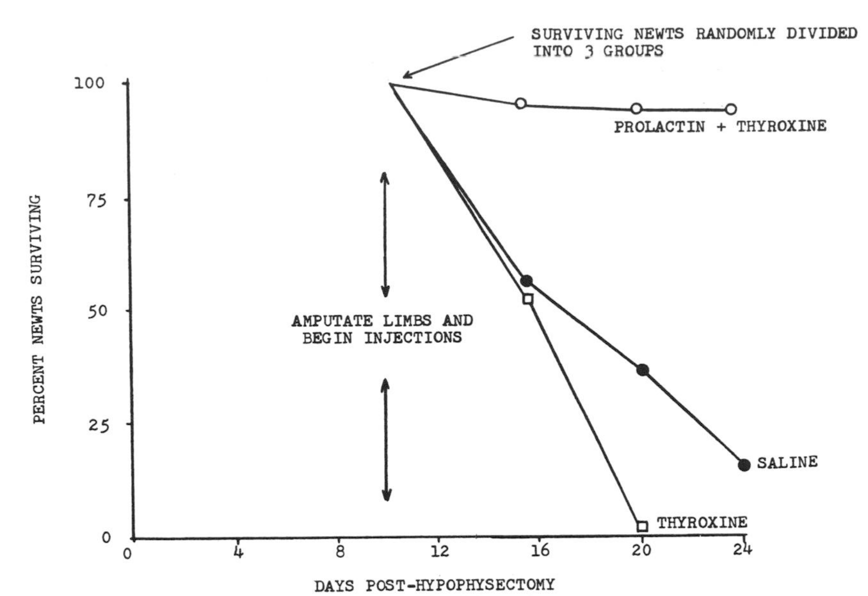

Fig. 6 A comparison of the per cent survival of hypophysectomized newts treated with prolactin (0.015 U/newt/2 days) + thyroxine (1×10^{-7} conc.), thyroxine alone (1×10^{-7} conc.) and 0.9% saline (0.1 cm^3/newt/2 days).

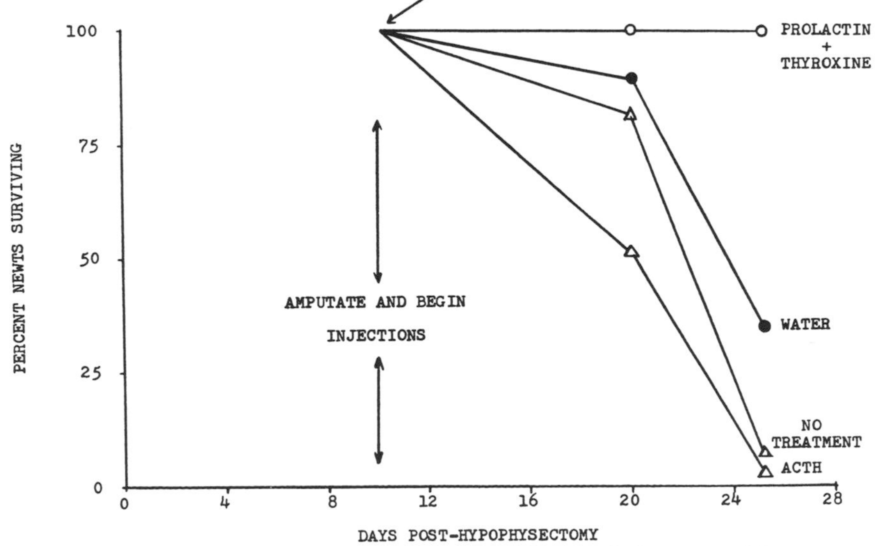

Fig. 7 A comparison of the per cent survival of hypophysectomized newts receiving no treatment, prolactin (0.015 U/newt/2 days) + thyroxine (1×10^{-7} conc.), ACTH (1 U/newt/2 days), and water (0.1 cm^3/newt/2 days).

TABLE 5

Hormone quantities injected into hypophysectomized newts

Preparation	Hormone activities present	Amount injected/newt every other day	
		Dose 1	Dose 2
Growth hormone Bovine NIH-GH-B12	Growth hormone (GH) 1.0 U/mg	0.3 mg = 0.291 U	0.03 mg = 0.0291 U
	Thyroid stimulating Hormone (TSH) 0.013 U/mg	0.0039 U	0.00039 U
	Luteinizing hormone (LH) 0.0085 U/mg	0.00255 U	0.000255 U
	Follicle stimulating Hormone (FSH) 0.001 U/mg	0.0003 U	0.00003 U
	Prolactin 0.54 U/mg	0.162 U	0.0162 U
Prolactin Bovine NIH-P-B2	Prolactin 19.9 U/mg	0.015 U	
	TSH 0.0002 U/mg	0.00000015 U	
	GH 0.01 U/mg	0.0000075 U = 0.00075 mg	
	LH 0.0004 U/mg	0.0000003 U	
	FSH 0.01 U/mg	0.0000075 U	
ACTH UpJohn Co.	ACTH 46.3 U/mg (No contaminants listed)	1.0 U	

fasted for two weeks prior to hypophysectomy.

Histological examination of a total of over 40 heads (Series I, II, and III) revealed complete hypophysectomies in over 97% of the cases (fig. 11). In addition, the ectopic pituitary grafts from both newts and axolotls appeared healthy and well vascularized even 35 days after grafting (fig. 13). No neural lobe tissue was observed in these grafts. The skin of the hypophysectomized newts with ectopic pituitary grafts became dark due to excess MSH secretion from the ectopic gland (Tassava et al., '68). The skin of these newts was smooth and slippery and not rough and black as is typical of the skin of untreated hypophysectomized adult newts. In addition, hypophysectomized newts treated with prolactin + thyroxine, GH, or ectopic pituitary grafts were active and readily ate fresh beef liver 30 days post-hypophysectomy. They also underwent occasional molts. Limb regeneration was normal in this group of newts (tables 3, 4) whereas, hypophysectomized newts treated with saline, prolactin alone, ACTH and/or thyroxine exhibited delayed limb regeneration (tables 3, 4).

The *intact* newts injected with ACTH (UpJohn) became dark in color and remained dark during the 24 days of treatment, whereas the eight intact newts injected with water appeared normal. All eight newts in each group survived the experiment and newts of both groups regenerated their amputated limbs in a comparable manner; 100% of the amputated limbs of the 16 newts showed typical regeneration on day 20 post-amputation. On day 30 post-amputation, all 16 limbs had reached the digit stage of regeneration (see footnote, table 1).

In summary, the results of Series III indicate that prolactin + thyroxine, growth hormone, and ectopic pituitary grafts effectively enhance both survival and limb regeneration of hypophysectomized newts. ACTH had no significant effect either on survival or limb regeneration of either hypophysectomized or intact newts.

DISCUSSION

The results described in the preceding pages make it clear that adult newt limb regeneration can be initiated and will proceed to at least the medium bud stage in the complete absence of pituitary hormones. Thus, the conclusion of Schotté ('61), that the stress of limb amputation initiates pituitary ACTH secretion and consequent release of adrenal steroids which are essential to proper wound healing, is not supported by the above data. These data are consistent with the findings of Liversage ('59), however, that amputated limbs will regenerate even though previously isolated from the central nervous system by spinal cord ablation.

It was apparent that regenerates on normal intact newts were more advanced in development than those on hypophysectomized newts of a comparable age, therefore indicating that pituitary hormones insure a more rapid rate of regeneration. These results agree with the observations of DeConinck et al. ('55) who found that hypophysectomized *Triturus alpestris* and *Triturus vulgaris* regenerated limbs which were amputated simultaneously with or four days after hypophysectomy; however, regeneration was considerably delayed. Hay ('56) and DeConinck et al. ('55) suggested that the primary influence of pituitary hormones was on the growth of the regeneration blastema and the results of the present study support this credence. Regenerates which had developed for 14 days in the presence of pituitary hormones nevertheless required these pituitary hormones to continue *normal* development. The 14 day regenerates increased in size after hypophysectomy but to a significantly lesser degree than did those of their sham-operated counterparts. Schotté and Hall ('52) also found that pituitary hormones were necessary for the normal growth of the limb blastema. By delaying hypophysectomy until 14 days after limb amputation, these authors obtained abortive or delayed regenerates in over 75% of the cases. Even more adverse effects were noted when hypophysectomy was delayed for only 7–13 days after limb amputation; 100% of the regenerates were abortive or delayed (Schotté and Hall, '52). The blastema cells begin DNA synthesis as early as the fourth day after amputation (Hay and Fischman, '61) and cell proliferation is an essential part of blastema formation (Chalkley, '54). Therefore it is not surprising that limb regeneration seldom proceeds to completion when hypophysectomy is performed prior to, concomitantly with, or shortly after limb amputation (Hall and Schotté, '51; Schotté and Hall, '52; Connelly et al., '68).

In addition to being able to regenerate lost limbs, the adult newt can readily replace an excised lens. The hormonal requirement is apparently comparable for regeneration of both organs. This supposition is supported by the data of the present investigation and by the experiments of Stone and Steinitz ('53) who found that when newts were lentectomized 5–8 days after hypophysectomy, regeneration was considerably delayed and the final result was often an abnormal lens. Similar delays and abnormalities of lens regeneration were observed in thyroidectomized newts. Thus, in the absence of the pituitary and/or the thyroid, there occurs a retardation in the rate of development of the regenerate of the limb and of the lens, both accompanied by various abnormalities. It is important to note that in the present experiments and in the experiments of Stone and Steinitz ('53), regeneration could not be correlated with small pituitary remnants due to improper hypophysectomies. Thus, thyroid and pituitary hormones are essential to both *normal* lens regeneration (Stone and Steinitz, '53) and *normal* limb regeneration (Connelly et al., '68). Hormones from the *in situ* pituitary, without thyroid hormone, are not enough to promote normal limb or lens regeneration (Richardson, '45; Schotté and Washburn, '54; Stone and Steinitz, '53). Connelly et al. ('68) found that hypophysectomized newts treated with prolactin alone (1.2 U/newt every other day) survived significantly longer than hypophysectomized saline-treated newts but not as long as newts treated with prolactin + thyroxine. The hypophysectomized newts which survived when treated with the larger amount of prolactin all regenerated amputated limbs. The large dose of prolactin (NIH) used may have contained enough TSH as a contaminant to activate the thyroid to a minimum level. It may be that very little thyroxine is needed for the normal health of newts when combined with prolactin, since as the data of the present study demonstrate, even a 1×10^{-8} concentration of thyroxine, with prolactin, significantly enhances survival and limb regeneration of hypophysectomized newts. Of interest in this connection are the experiments of Schmidt ('58) who found that hypothyroid newts regenerated limbs even faster than normal newts. A low thyroxine level may increase prolactin secretion and the higher prolactin quantity may then stimulate regeneration (see also Thornton, '68). Grant and Cooper ('65) have shown that when thyroxine levels are low, as in the adult newt, prolactin secretion is high (aquatic environment), whereas when thyroxine levels are high, as in the eft, prolactin levels are low (land environment). These authors also report that thyroxine treatment will cause aquatic newts to resume a land habitat, again suggesting decreased prolactin secretion under conditions of high thyroxine. High exogenous thyroxine will cause abnormal limb regeneration of intact newts and the effect of the thyroxine appears to be most pronounced during the growth phase, beginning two weeks after amputation (Hay, '56). Since it was shown in this investigation that pituitary hormones, specifically prolactin and TSH, are important to the normal growth of the blastema, it is tempting to speculate that in Hay's ('56) experiments the high thyroxine levels decreased prolactin secretion by the pituitary, thus resulting in a type of *in situ* hypophysectomy.

In the present study, a prolactin-thyroxine combination was found to be very effective in enhancing survival and also limb regeneration of hypophysectomized newts. This hormone combination resulted in survival of almost 100% of the hypophysectomized newts so treated, even when the hormones were not administered until ten days post-hypophysectomy. Thyroxine alone in the concentration used was ineffective in enhancing survival and limb regeneration and prolactin alone (0.015 U/newt every other day) was also ineffective. These findings strongly suggest that prolactin and thyroxine act synergistically in some as yet unknown way and the combination of the two hormones is essential to normal health and limb regeneration of the adult newt. Growth hormone was shown to be as effective as the prolactin-thyroxine combination in enhancing survival and limb regeneration of hypophysectomized newts, even at 1/10 the quantity used by Wilkerson ('63). The effectiveness of this smaller quantity of growth hormone (NIH) used in the present investigation may have been due to the fact that it contained ten times as much prolactin per mg of preparation as the growth hormone (NIH) used by Wilkerson. Thus, the newts given 0.03 mg of growth hormone received 0.016 U prolactin/newt every other day. It was pointed out by Berman et al. ('64) that this amount of prolactin, equivalent to the prolactin contaminant which Wilkerson ('63) administered to newts, will elicit a growth response when injected into frog tadpoles. It cannot be said with certainty whether the quantity of growth hormone administered in this investigation contained enough TSH to activate the thyroid of a hypophysectomized newt. However, occasional molting or partial molting was observed during the treatment period suggesting some thyroid activation. Richardson ('45) and Wilkerson ('63) both suggested that some thyroid activity is important to normal limb regeneration. Additional investigations should determine whether growth hormone, completely free of prolactin and/or TSH would still be effective in supporting limb regeneration and survival of hypophysectomized newts.

The finding that a prolactin-thyroxine combination is important to newt limb regeneration and survival agrees with the abundant evidence that prolactin and TSH are produced by the newt pituitary and that these hormones (TSH acting via the thyroid) may influence newt physiology. The larval newt undergoes a primary metamorphosis to a land form, called an eft, which lives from 1–5 years on land before migrating to water as an aquatic, reproductively mature adult. The "water drive" of the eft has been shown to be induced by prolactin and the migratory behavior, and the structural and physiological changes involved, have been termed "second metamorphosis" (Grant, '61). The induced water drive of the eft has been recommended as an adequate assay for prolactin (Grant, '59) and has been used to demonstrate the presence of prolactin in pituitary tissue of *Bufo, Fundulus, Cyprinus, Natrix* (Chadwick, '41; Grant, '61) and adult newts (Reinke and Chadwick, '39). Antuitrin G will induce water drive (Chadwick, '40) but LH, ACTH, posterior pituitary tissue, TSH, and Antuitrin S have no water drive activity (Grant and Grant, '58). Since the ectopically transplanted eft pituitary will induce a water drive (Masur, '62) it is probable that prolactin secretion is under negative control by the hypothalamus (Grant, '61). The ectopic pituitary of the adult newt has been shown to produce thyroid stimulating hormone (TSH) in near normal amounts (Dent, '66) and ultrastructural studies of the adult newt pituitary suggest that TSH and prolactin are produced by the normal as well as the ectopic gland (Dent and Gupta, '67).

There is no evidence that prolactin and/or thyroxine act through the adrenal gland thereby increasing or maintaining the secretion of adrenal hormones. Adrenal hormones did not enhance survival of hypophysectomized newts (Schotté and Bierman, '56) and in this investigation prolactin and thyroxine, but not ACTH, significantly enhanced survival and limb regeneration of hypophysectomized newts. Furthermore, in mammals prolactin causes a decreased secretion of the adrenal gland (Bates et al., '64). Why then, was the ACTH used by Schotté and Chamberlain ('55) beneficial to survival and limb regeneration of hypophysectomized newts? It is likely that ACTH was *not* the active hormone in the preparation used by Schotté and Chamberlain ('55) for three reasons: (1) their ACTH contained 1.14 U/ACTH/mg (crude Armour ACTH, '55 preparation) whereas it was not until 1962 that Armour produced purified ACTH containing 33 U/ACTH/mg (Evans et al., '66); the ACTH used in this investigation, on the other hand, was essentially pure ACTH (UpJohn Co., '66 preparation). (2) The description given by Schotté and Chamberlain ('55) of the response of the hypophysectomized newts to the crude ACTH — "newts remained active, had good appetites, and slippery skin" — closely resembles the response of the hypophysectomized newts in this investigation to prolactin + thyroxine, growth hormone, or ectopic pituitary grafts. In the present investigation it was found that hypophysectomized newts given pure ACTH (UpJohn) are sluggish, will not feed, and instead of having slippery skin, have course granular skin similar to that of saline-treated hypophysectomized newts. (3) Schotté and Chamberlain ('55) reported that ACTH (1 U/newt/2 days) resulted in inhibition of limb regeneration in intact newts. This effect could also have been due to contaminating hormones. In this investigation, pure ACTH (1 U/newt/2 days) *did not* inhibit limb regeneration in intact newts. Therefore, the evidence implies that prolactin and thyroxine are essential to adult newt survival and normal limb regeneration and furthermore, that adrenal hormones are not limiting after hypophysectomy.

The exact pathway by which prolactin and thyroxine influence survival and limb regeneration is unknown and more work must be done on this problem. However, some possible functions of prolactin and thyroxine in the newt are suggested by observations of the present and other investigations. Prolactin may act on cell proliferation. Inoue ('56) found a diminished mitotic proliferation in epidermal cells of amputated limbs of hypophysectomized adult newts. Waterman ('65) and Niwelinski ('58) increased the rate of *intact* newt limb regeneration by prolactin treatment. Waterman ('65) also found that prolactin increased appetite and body weight of intact newts. Thyroxine will act directly on the skin of adult newts (Clark and Kaltenbach, '61) and also on denervated skin (Taban and Tassava, unpublished) and Grant and Cooper ('65) found that prolactin would maintain newt skin in organ culture but thyroxine alone was ineffective. Prolactin also acts on the skin of lizards by raising the frequency of sloughing (Maderson and Licht, '67) and prolactin treatment will increase the molting frequency and the mitotic rate of red eft skin (Chadwick and Jackson, '48).

Grant ('61) suggested that prolactin may induce the water drive of the red eft because of the role of this hormone in the water balance of the newt. This suggestion is important since it is known that prolactin does influence water balance in eels (Olivereau and Ball, '64), *Fundulus* (Ball and Ensor, '67) and *Tilapia* (Dharmamba et al., '67). In this investigation, survival of hypophysectomized newts was not enhanced by maintaining newts in NaCl solutions; however, measurements of tissue sodium and sodium transport before and after hypophysectomy would be a more worthwhile approach to this problem.

Prolactin may also play a role in metabolism of fat, protein and carbohydrate. Hypophysectomized *Tilapia mossambica* cannot form liver glycogen from amino acid precursors (Swallows and Fleming, '67) which suggests that pituitary hormones are essential for gluconeogenesis in these fish. Prolactin enhances food consumption and body weight gain in both newts (Waterman, '65) and lizards (Licht, '67) and prolactin treated lizards also show a significant weight increase of regenerating tails. It may be that prolactin and thyroxine are involved in normal metabolism in the adult newt and these hormones are required for energy (glucose) production from protein. Thus, hypophysectomized newts which were previously fasted, survived only when given prolactin + thyroxine. Furthermore, hypophysectomized newts which were previously well fed survived significantly longer than fasted hypophysectomized newts. These observations suggest that newts in particularly good nutritional state at the time of hypophysectomy may contain food reserves, such as liver glycogen, which can be utilized for energy in the absence of hormones. The energy from the food reserves could then be used to maintain survival and also for growth of the regenerate. This speculation is supported by the fact that newts in Series I which were fed daily for two weeks weighed significantly more at the time of hypophysectomy and survived significantly longer after hypophysectomy than newts fasted for two weeks prior to hypophysectomy. Whether prolactin and thyroxine act on metabolism, which cells and tissues are acted upon and exactly how these hormones interact in enhancing survival and limb regeneration of adult newts will receive further attention in future investigations.

Schotté ('61) reported that larval *Ambystoma punctatum* and larval newt pituitary grafts did not support limb regeneration in hypophysectomized adult newts. It was therefore surprising to discover in this investigation that larval *Ambystoma mexicanum* (axolotl) pituitaries (2 pituitary grafts/newt) did significantly enhance survival and limb regeneration of hypophysectomized newts. Prolactin has been identified in the pituitary of *Necturus* and adult *Ambystoma tigrinum* (Nicoll and Bern, '68) and the results of the present experiments suggest that larval *Ambystoma mexicanum* pituitary tissue also contains prolactin. The epidermis of hypophysectomized newts with ectopic axolotl pituitary grafts does not build up as in hypophysectomized newts given only prolactin; and, in addition, these newts undergo occasional molts. Thus, the axolotl pituitary, ectopically transplanted to the newt apparently also secretes TSH, thus activating the newt's thyroid. It may be that in the axolotl no TSH releasing factor is present, whereas, grafted in the newt, the axolotl pituitary responds to TSH releasing factor which reaches the pituitary through the blood circulation.

ACKNOWLEDGMENTS

This investigation was conducted while I held a predoctoral fellowship (NDEA) and was included in a dissertation submitted to Michigan State University in partial fulfillment of the requirements for the Ph.D., June, 1968. This work was supported by NIH grant NB-04128 and NSF grants GE-2618 and GB-7748 administered by Dr. C. S. Thornton, Department of Zoology, Michigan State University. I would like to express my appreciation to Dr. Thornton for his helpful suggestions and advice during the course of this investigation and in the preparation of the manuscript.

LITERATURE CITED

Ball, J. N., and D. M. Ensor 1967 Specific action of prolactin on plasma sodium levels in hypophysectomized *Poecilia latipinna* (Teleostei). Gen. Comp. Endo., *8(3)*: 432–440.

Bates, R. W., S. Milkovic and M. M. Garrison 1964 Effects of prolactin, growth hormone, and ACTH, alone and in combination, upon organ weights and adrenal function in normal rats. Endo., *74*: 714–723.

Berman, R., H. A. Bern, C. S. Nicoll and R. C. Strohman 1964 Growth promoting effects of mammalian prolactin and growth hormone in

tadpoles of *Rana catesbeiana.* J. Exp. Zool., *156:* 353–360.
Bragdon, D. E., and J. N. Dent 1954 Effect of ACTH and cortisone on renal fat and limb regeneration in adult salamanders. Proc. Soc. Exp. Biol. Med., *87:* 460–462.
Chadwick, C. S. 1940 Induction of water drive in *Triturus viridescens* with anterior pituitary extract. Proc. Soc. Exp. Biol. Med., *43:* 509–511.
——— 1941 Further observations on the water drive in *Triturus viridescens.* II. Induction of the water drive with the lactogenic hormone. J. Exp. Zool., *86:* 175–187.
Chadwick, C. S., and H. R. Jackson 1948 Acceleration of skin growth and molting in the red eft of *Triturus viridescens.* Anat. Rec., *101:* 718.
Chalkley, D. T. 1954 A quantitative histological analysis of forelimb regeneration in *Triturus viridescens.* J. Morph., *94:* 21–70.
Clark, N. B., and J. C. Kaltenbach 1961 Direct action of thyroxine on the skin of the adult newt. Gen. Comp. Endo., *1:* 513–518.
Connelly, T. G., R. Tassava and C. S. Thornton 1968 Survival and limb regeneration of prolactin treated hypophysectomized newts. J. Morph. In press.
DeConinck, L., M. Denuce, Fr. Dierckx and M. Janssens 1955 Acides ribonucleiques, hormone somatotrope et regeneration chez *Triturus* (Urodeles). Societe Royale Zoologique De Belgique Annales, *LXXXVi:* 191–234.
Dent, J. N. 1966 Maintenance of thyroid function in newts with transplanted pituitary glands. Gen. Comp. Endo., *6:* 401–408.
——— 1967 Survival and function in hypophysial homografts in the spotted newt. Amer. Zool., *7(4):* 714.
Dent, J. N., and B. J. Gupta 1967 Ultrastructural observations on the developmental cytology of the pituitary gland in the spotted newt. Gen. Comp. Endo., *8:* 273–288.
Dharmamba, R., R. I. Handin, J. Nandi and H. A. Bern 1967 Effect of prolactin on fresh water survival and on plasma osmotic pressure of hypophysectomized *Tilapia mossambica.* Gen. Comp. Endo., *9:* 295–302.
Evans, H. M., L. L. Sparks and J. S. Dixen 1966 The physiology and chemistry of adrenocorticotropin. In: The Pituitary Gland. Vol. 1. G. W. Harris and B. T. Donovan, eds. Univ. Calif. Press, Berkeley.
Fortier, C. 1966 Nervous control of ACTH secretion. In: The Pituitary Gland. Vol. 2. G. W. Harris and B. T. Donovan, eds. Univ. Calif. Press, Berkeley.
Geschwind, I. I. 1967 Molecular variation and possible lines of evolution of peptide and protein hormones. Amer. Zool., *7:* 89–108.
Grant, W. C. 1959 A test for prolactin using the hypophysectomized eft stage of *Diemictylus viridescens.* Endo., *64:* 839–841.
——— 1961 Special aspects of the metamorphic process: second metamorphosis. Amer. Zool., *1:* 163–171.
Grant, W. C., and J. A. Grant 1958 Water drive studies on hypophysectomized efts of *Diemictylus viridescens.* Part I. The role of the lactogenic hormone. Biol. Bull., *114(1):* 1–9.
Grant, W. C., and G. Cooper 1965 Behaviorial and integumentary changes associated with induced metamorphosis in *Diemictylus.* Biol. Bull., *129:* 510–522.
Hall, A. B., and O. E. Schotté 1951 Effects of hypophysectomies upon the initiation of regenerative processes in the limb of *Triturus viridescens.* J. Exp. Zool., *118:* 363–382.
Hanke, W., and K. Weber 1965 Histophysiological investigation on the zonation, activity, and mode of secretion of the adrenal gland of the frog, *Rana temporaria.* Gen. Comp. Endo., *5:* 444–455.
Hay, E. D. 1956 Effects of thyroxine on limb regeneration in the newt *Triturus viridescens.* Bull. Johns Hopkins Hosp., *99:* 262–285.
——— 1966 Regeneration. Holt, Rinehart and Winston, N. Y., N. Y.
Hay, E. D., and D. A. Fischman 1961 Origin of the blastema in regenerating limbs of the newt *Triturus viridescens.* Dev. Biol., *3:* 327–342.
Inoue, S. 1956 Effect of growth hormone and cortisone acetate upon mitotic activity in normal and regenerating tissues of amphibians. Endo. Japan., *3:* 236–239.
Licht, P. 1967 Interaction of prolactin and gonadotropins on appetite, growth and tail regeneration in the lizard, *Anolis carolinensis.* Gen. Comp. Endo., *9(1):* 49–63.
Liversage, R. A. 1959 The relation of the central and autonomic nervous systems to the regeneration of limbs in adult urodeles. J. Exp. Zool., *141:* 75–118.
Maderson, P. F. A., and P. Licht 1967 Epidermal morphology and sloughing frequency in normal and prolactin treated *Anolis carolinensis.* J. Morph., *123:* 157–172.
Mangili, G., M. Motta and L. Martini 1966 Control of adrenocorticotropic hormone secretion. In: Neuroendocrinology. Vol. 1. L. Martini and W. F. Ganong, eds. Academic Press, N. Y.
Masur, S. 1962 Autotransplantation of the pituitary of the red eft. Amer. Zool., *2:* 538.
Merchant, D. J., R. H. Kahn and W. H. Murphy 1964 Handbook of Cell and Organ Culture. Burgess Publishing Co., Minneapolis, Minn.
Nicoll, C. S., and H. A. Bern 1968 Further analysis of the occurrence of pigeon crop-stimulating activity (prolactin) in the vertebrate hypophysis. Gen. Comp. Endo., *11(1):* 5–20.
Niwelinski, J. 1958 The effect of prolactin and somatotropin on the regeneration of the forelimb in the newt, *Triturus alpestris.* Folia Biol., *6:* 9–36.
Olivereau, M., and J. N. Ball 1964 A contribution to the histophysiology of the hypophysis of teleosteans in particular the cells of *Poecillia.* Gen. Comp. Endo., *4:* 523–532.
Pearse, A. G. E. 1961 Histochemistry: Theoretical and Applied. Little, Brown and Co., Boston.
Piper, G. D., and R. DeRoos 1967 Evidence for a corticoid-pituitary negative feedback mech-

anism in the American bullfrog (*Rana catesbeiana*). Gen. Comp. Endo., *8:* 135–142.

Purves, H. D., and N. E. Sirett 1967 Corticotropin secretion by ectopic pituitary glands. Endocrinology, *30:* 962–968.

Reinke, E. E., and C. S. Chadwick 1939 Inducing land stage of *Triturus viridescens* to assume water habitat by pituitary implantation. Proc. Soc. Exp. Biol. Med., *40:* 671–693.

Richardson, D. 1945 Thyroid and pituitary hormones in relation to regeneration. II. Regeneration of the hind limb of the newt, *Triturus viridescens*, with different combinations of thyroid and pituitary hormones. J. Exp. Zool., *100:* 417–429.

Schmidt, A. J. 1958 Forelimb regeneration of thyroidectomized adult newts. II. Histology. J. Exp. Zool., *139:* 95–125.

Schotté, O. E. 1961 Systemic factors in initiation of regenerative processes in limbs of larval and adult amphibians. In: Molecular and Cellular Structures, 19th Growth Symposium. D. Rudnick, ed. The Ronald Press Co., N. Y.

Schotté, O. E., and R. H. Bierman 1956 Effects of cortisone and allied steroids upon limb regeneration in hypophysectomized *Triturus viridescens*. Rev. Suisse Zool., *63:* 353–375.

Schotté, O. E., and J. L. Chamberlain 1955 Effects of ACTH upon limb regeneration in normal and in hypophysectomized *Triturus viridescens*. Rev. Suisse Zool., *62:* 253–279.

Schotté, O. E., and A. B. Hall 1952 Effect of hypophysectomy upon phases of regeneration in progress, *Triturus viridescens*. J. Exp. Zool., *121:* 521–556.

Schotté, O. E., and D. A. Lindberg 1954 Effect of xenoplastic adrenal transplants upon limb regeneration in normal and in hypophysectomized newts (*Triturus viridescens*). Proc. Soc. Exp. Biol. Med., *87:* 26–29.

Schotté, O. E., and A. Tallon 1960 The importance of autoplastically transplanted pituitaries for survival and for regeneration of adult *Triturus*. Experimentia, *16:* 72–76.

Schotté, O. E., and W. W. Washburn 1954 Effect of thyroidectomy on the regeneration of the forelimb in *Triturus viridescens*. Anat. Rec., *120:* 156.

Singer, M. 1952 The influence of the nerves in regeneration of the amphibian extremity. Quart. Rev. Biol., *27(2):* 169–200.

Stone, L. S., and H. Steinitz 1953 Effects of hypophysectomy and thyroidectomy on lens and retina regeneration in the adult newt, *Triturus v. viridescens*. J. Exp. Zool., *124:* 469–504.

Swallows, R. L., and W. R. Fleming 1967 Effect of hypophysectomy on the metabolism of liver glycogen of *Tilapia mossambica*. Amer. Zool., *7(4):* 715.

Tassava, R. A., F. J. Chlapowski and C. S. Thornton 1968 Limb regeneration in *Ambystoma* larvae during and after treatment with adult pituitary hormones. J. Exp. Zool., *167:* 157–163.

Thornton, C. S. 1968 Amphibian limb regeneration. In: Advances in Morphogenesis. Vol. 7. Academic Press, N. Y.

Van Dongen, W. J., D. B. Jorgensen, L. O. Larsen, P. Rosenkilde, B. Lofts and P. G. W. J. Van Oordt 1966 Function and cytology of the normal and autotransplanted pars distalis of the hypophysis in the toad, *Bufo bufo*. Gen. Comp. Endo., *6:* 491–518.

Waterman, A. J. 1965 Prolactin and regeneration of the forelimbs of the newt. Amer. Zool., *5:* 237.

Wilkerson, J. A. 1963 The role of growth hormone in regeneration of the forelimb of the hypophysectomized newt. J. Exp. Zool., *154:* 223–230.

Wurster, D. H., and M. R. Miller 1960 Studies on the blood glucose and pancreatic islets of the salamander, *Taricha torosa*. Comp. Biochem. Physiol., *1:* 101–109.

PLATE 1

EXPLANATION OF FIGURES

8 A longitudinal section through the regeneration blastema of a hypophysectomized newt which was fed for two weeks prior to hypophysectomy. This limb was amputated five days post-hypophysectomy and fixed 26 days post-hypophysectomy. Histological examination of the head of this newt demonstrated complete hypophysectomy. Masson's Trichrome Stain. × 160.

9 A longitudinal section through the regeneration blastema of a hypophysectomized newt which was fasted for two weeks prior to hypophysectomy. This limb was amputated five days post-hypophysectomy and fixed 26 days post-hypophysectomy. Hematoxylin-eosin. × 160. Enlarged × 1.5.

10 A median sagittal section through the pituitary gland of a sham-operated newt. PAS Stain. × 100.

11 A median sagittal section through the base of the infundibulum of a completely hypophysectomized newt which was fed for two weeks prior to hypophysectomy. At the time of fixation, 26 days post-hypophysectomy, both forelimbs of this newt showed positive regeneration. PAS Stain. × 100.

12 A median sagittal section through the base of the infundibulum of a partially hypophysectomized newt which was fasted for two weeks prior to hypophysectomy. The small pituitary fragment (arrow) had little or no survival value since this newt survived for only 17 days post-hypophysectomy. PAS Stain. × 100.

13 A median sagittal section through the grafted larval axolotl (*A. mexicanum*) pituitary tissue. This completely hypophysectomized newt was fixed 35 days post-hypophysectomy. At the time of fixation (30 days post-amputation) the forelimbs of this newt exhibited three digit regenerates. Herlant's Tetrachrome Stain. × 100.

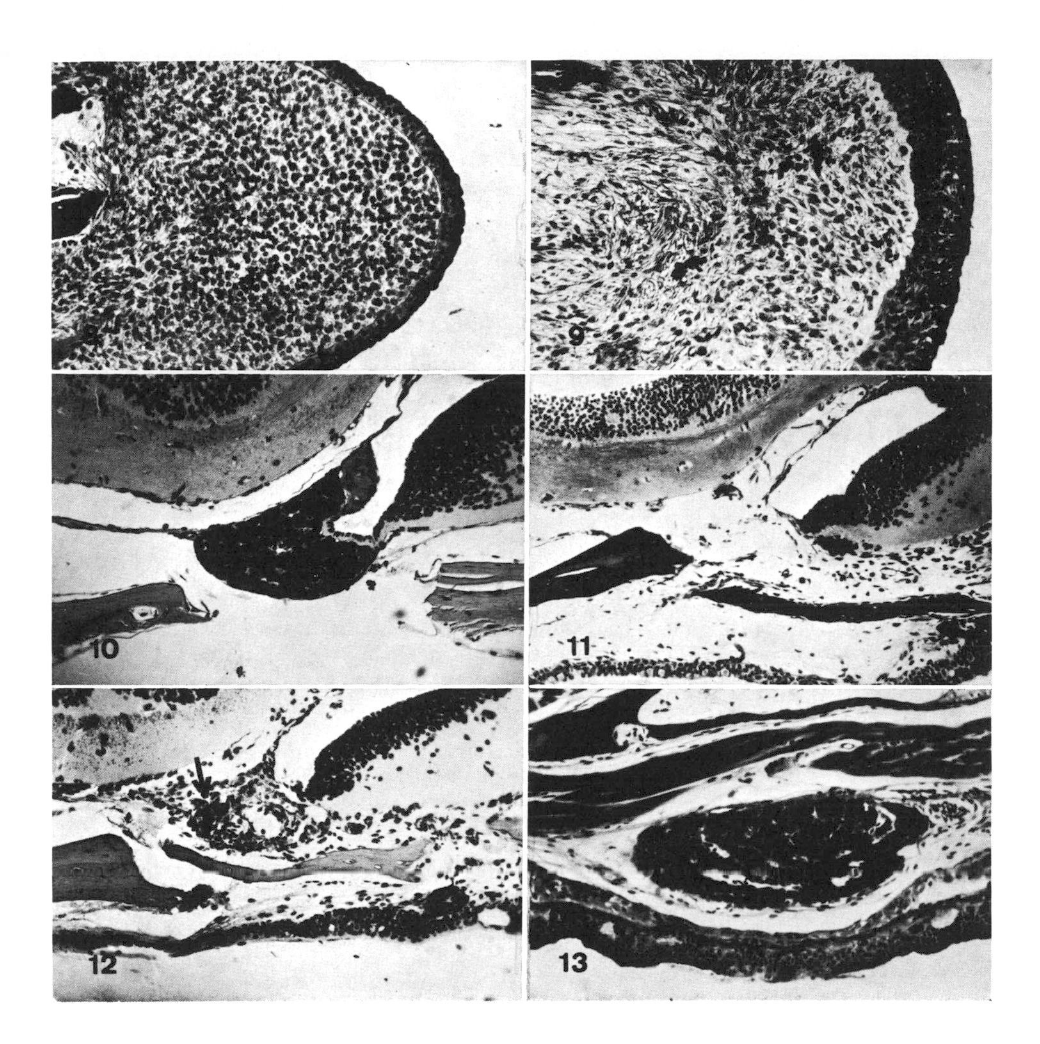
8
9
10
11
12
13

Survival and Limb Regeneration of Hypophysectomized Newts with Pituitary Xenografts from Larval Axolotls, *Ambystoma mexicanum*

ROY A. TASSAVA

Hypophysectomized adult newts, *Notophthalmus viridescens,* which normally die within four weeks, will survive for over four months when given adult newt pituitary homografts (Dent, '67; Schotté and Tallon, '60). These ectopic glands apparently secrete thyroid stimulating hormone (TSH) since the normal activity of the thyroid is maintained and the epidermis of the skin does not build up (Dent, '66). Although there are conflicting reports (Tassava, '68), Schotté ('61) originally proposed that the pituitary gland is indispensable for adult newt limb regeneration and reported that an ectopic *adult* newt pituitary would support limb regeneration, but an ectopic larval newt or larval *Ambystoma* pituitary was ineffective. Since larval *Ambystoma* do not require their pituitary gland for limb regeneration (Liversage, '67; Tassava *et al.*, '68), Schotté ('61) suggested that the larval pituitary gland does not secrete the hormones necessary for limb regeneration of the adult newt.

In the present report, the hypophysectomized adult newt served as a test system to determine the nature of the hormones secreted by the pituitary of larval axolotls, *Ambystoma mexicanum*. Three questions were asked: (1) Will the axolotl pituitary secrete TSH and activate the thyroid of hypophysectomized adult newts? (2) Will hypophysectomized adult newts with ectopic larval *Ambystoma mexicanum* pituitaries regenerate amputated limbs? (3) Will the pituitary xenograft be rejected and if so, how soon?

MATERIALS AND METHODS

Adult newts were obtained from Lewis Babbitt, Petersham, Massachusetts, and were fasted for two weeks prior to hypophysectomy. Larval axolotls were a gift from Dr. R. R. Humphrey, Department of Zoology, University of Indiana. After anesthetization in MS 222 a small piece of bone was cut free immediately ventral to the pituitary of 21 newts and the whole gland was removed by suction with a fine glass pipette (Connelly *et al.*, '68). The bone piece was then replaced. The entire pituitary gland of each of 15 larval axolotls was next inserted into a small tunnel under the skin of the lower jaw of each of 15 of the newts. The snout-tail tip length of the axolotls ranged from 10 to 16 cm. The remaining 6 hypophysectomized newts were implanted in the same manner with axolotl brain tissue. Since hypophysectomized newts stop feeding, none of the newts were fed for the first 30 days and from then on the survivors were fed beef liver every three days.

Five days after hypophysectomy and pituitary grafting, both forelimbs of each newt were amputated through the radius and ulna. The length of survival, skin color and texture and status of limb regeneration of each newt were observed at regular intervals. At the time of death, heads and limbs were fixed and prepared for histology by the method of Stone ('66). Sections of heads, cut at 10 μ, were stained with the PAS or Herlant's tetrachrome techniques (Pearse, '61). Limb sections were stained with hematoxylin and eosin.

To test whether survival was due to a functioning xenograft or to pituitary remnants left after faulty hypophysectomies, the xenografts were removed from five of the ten surviving newts 146 days after hypophysectomy, and forelimbs were again amputated. The skin texture, length of survival and amount of limb regeneration of the newts were again observed. These long term xenografts will be studied by differential staining and electron microscopy (Connelly *et al.*, '68). The other five surviving newts were sham-operated by cutting into the lower jaw without removing the xenograft. The forelimbs of these five sham-operated newts were also amputated. The newts were not fed until 30 days after sham-operation.

RESULTS

The hypophysectomized newts became dark in color two days after grafting of the larval axolotl pituitary, apparently from melanophore stimulating hormone (MSH) secretion from the ectopic gland (fig. 1). The one newt which did not darken died six days after hypophysectomy and histology revealed that the pituitary graft had been lost. The skin of the hypophysectomized newts with grafts of axolotl brain tissue became rough and cornified (fig. 3). None of these six newts survived beyond 30 days after hypophysectomy (fig. 2) and limbs exhibited delayed regeneration. Hypophysectomized newts with pituitary xenografts appeared to have normal skin which was smooth in texture although dark in color (fig. 1). This normal skin condition was maintained as long as the ectopic gland was present (fig. 4).

The survival results are depicted in figure 2. The 14 newts with xenografts which survived to 18 days after limb amputation showed normal limb regeneration. When the 18 day regenerates were staged (see Tassava, '69), eight limbs were observed to possess medium bud blastemas, 11 possessed late bud blastemas, 7 exhibited palettes, and 2 were in digit stages.

Those newts which died during the course of the experiment, one each on days 26, 32, 81 and 101 after xenografting, were nevertheless dark at the time of death, suggesting that the MSH secreting portion of the gland was still functional.

Hypophysectomized newts become sluggish and normally do not eat even when repeatedly offered fresh beef liver (Tassava, '69). On the contrary, the hypophysectomized newts with axolotl pituitary xenografts readily ate beef liver and remained active during the time the xenograft was present.

The five newts whose pituitary xenografts were removed after 146 days became light in color within 12 hours, resembling normal newts, while the five sham-operated newts remained dark. Two-hundred days after grafting, two of the five sham-operated newts were decapitated and the xenograft was examined histologically. The xenografted glands appeared healthy and well vascularized. No brain tissue was observed in any of the grafts examined (fig. 5).

The skin of the five newts from which the pituitary grafts were removed became rough and cornified and these newts henceforth failed to molt (fig. 3). The skin of the sham-operated newts remained slippery and dark and molting was occasionally observed (fig. 4). None of the newts from which the grafts were removed survived beyond 25 days while four of the sham-operated newts (80%) were still surviving and appeared vigorous and in good health 60 days after the sham-operation which was over 200 days after grafting (fig. 2). These four newts regenerated amputated limbs in a typical fashion (fig. 6). Regeneration of the five newts from which the xenografts were removed was delayed (compare figs. 7 and 8). One xenografted newt was still surviving and in good health 270 days post-hypophysectomy. This newt was fixed and histology revealed complete hypophysectomy, normal limb regeneration and a healthy xenograft.

Fig. 1 A hypophysectomized newt (A) with an axolotl pituitary xenograft, 35 days after hypophysectomy. Note the dark, moist appearance of the skin. The larval axolotl (B) is typical of those used as pituitary donors in this investigation.

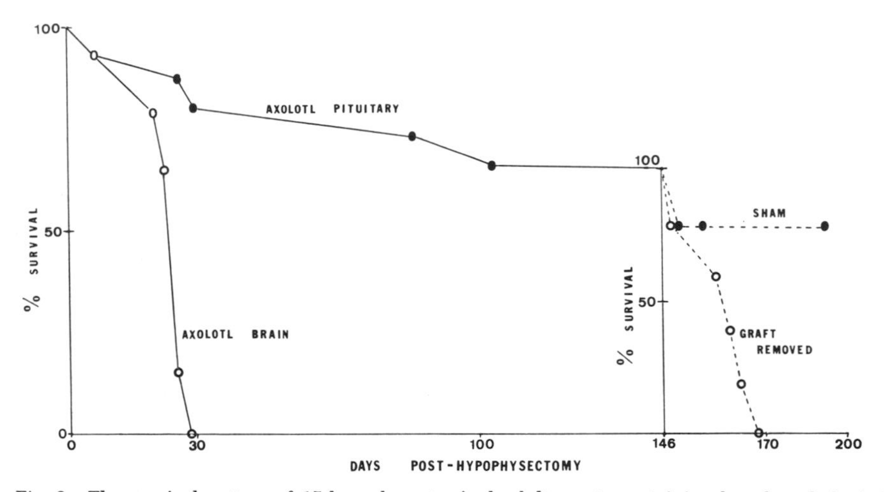

Fig. 2 The survival pattern of 15 hypophysectomized adult newts containing larval axolotl pituitary xenografts. Axolotl brain tissue was grafted to an additional 6 hypophysectomized newts. On day 146 post-hypophysectomy, the pituitary xenograft was removed from five of the ten surviving newts and the other 5 were sham-operated. Forelimbs were amputated 5 days after hypophysectomy and again on day 146. Hypophysectomized news with larval pituitary xenografts regenerated limbs in a normal fashion.

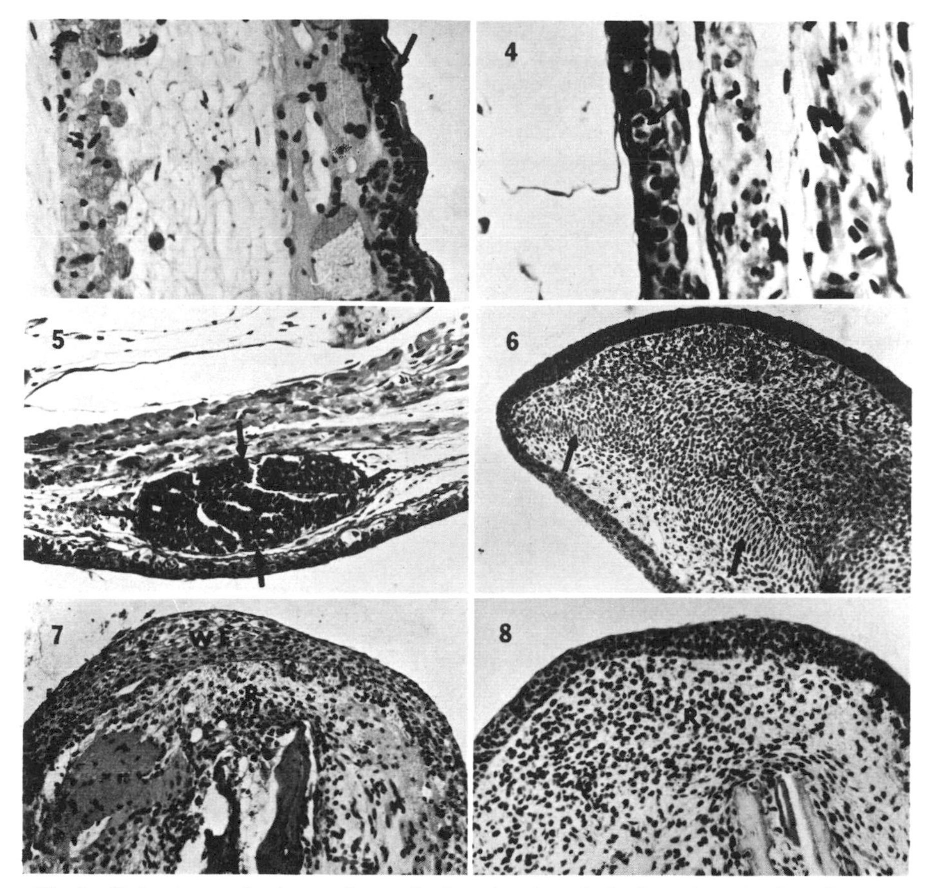

Fig. 3 Photomicrograph of a median sagittal section through the lower jaw of a hypophysectomized newt 16 days after removal of the larval axolotl pituitary xenograft. Note the lack of molting as evidenced by the thick, cornified layer of epithelial cells (arrow). Serial sections revealed no evidence of either newt or axolotl pituitary remnants in this head. Hematoxylin-eosin. 160 ×.

Fig. 4 Photomicrograph of a median sagittal section through the lower jaw of a hypophysectomized newt with a larval pituitary xenograft. This newt was sacrificed 200 days after grafting (54 days after sham-operation). Note normal skin with mitotic figure (arrow). Serial sections revealed complete hypophysectomy. Hematoxylin-eosin. 400 ×.

Fig. 5 Photomicrograph of a median sagittal section through the grafted larval axolotl (*A. mexicanum*) pituitary tissue (enclosed by arrows). This completely hypophysectomized newt was fixed 200 days after hypophysectomy. The forelimbs of this newt twice completed regeneration at a rate comparable to that of intact newts. Hematoxylin-eosin. 100 ×.

Fig. 6 Photomicrograph of a longitudinal section through a 28 day regeneration blastema of a completely hypophysectomized newt with a larval axolotl pituitary xenograft. This limb was amputated on the day of sham-operation, 146 days after grafting and hypophysectomy. Note the large regenerate and the beginning of cartilage differentiation (arrows). Hematoxylin-eosin. 100 ×.

Fig. 7 Photomicrograph of a longitudinal section through the regeneration blastema of a hypophysectomized newt which was fixed 16 days after amputation and xenograft removal. Note the thick wound epithelium (WE) but small regenerate (R) with few blastema cells. Hematoxylin-eosin. 100 ×.

Fig. 8 Photomicrograph of a longitudinal section through the 16 day regeneration blastema of a hypophysectomized newt with a pituitary xenograft. This limb was amputated on the day of sham-operation, 146 days after grafting and hypophysectomy. Note the normal, early bud regenerate (R). Hematoxylin-eosin. 100 ×.

DISCUSSION

Three observations in this report suggest that larval axolotl pituitaries are functional while grafted to hypophysectomized newts. (1) The newt skin darkened due to melanophore expansion and remained dark until the graft was removed. It is well known that ectopic pituitary glands, because they are removed from the MSH inhibiting factor of the hypothalamus, secrete an abundance of MSH (Geschwind, '67). (2) Histological examination of the heads of the newts with xenografts, which survived for over 146 days, revealed complete hypophysectomy in every case. Finally, (3) the skin of the hypophysectomized newts with pituitary xenografts showed occasional molting and remained smooth and slippery while the xenograft was present, but became rough and cornified shortly after graft removal. Furthermore, newts were active and ate readily while the xenograft was present whereas hypophysectomized newts typically become sluggish and refuse to eat (Tassava, '69).

The third observation noted above suggests that the larval axolotl pituitary, when grafted into the adult newt, secretes at least some TSH. This is a particularly interesting observation since the axolotl, *Ambystoma mexicanum*, normally does not undergo metamorphosis. Lynn and Wachowski ('51) suggested several possible defects which would lead to the condition of neoteny in amphibian larvae. (1) The tissues of the larvae could be unresponsive to thyroid hormone, (2) the thyroid could be defective and not produce thyroid hormone, (3) the pituitary could be defective and fail to produce thyroid stimulating hormone (TSH) to activate the thyroid, or finally, (4) the hypothalamus may not produce thyroid stimulating hormone releasing factor (TSHRF) so that the pituitary would not release TSH. The first of these possibilities was eliminated by Prahlad and DeLanney ('65) who obtained metamorphosis of larval axolotls by exogenous thyroxine. Thus the tissues of the axolotl will respond to exogenous thyroxine. There is also evidence that the thyroid of the axolotl is not at fault. Blount ('50) found that *A. mexicanum* larvae would metamorphose (2 cases) if given pituitary grafts from donor *A. tigrinum*. This finding suggests that the thyroid of the axolotl will respond to the proper TSH stimulus. However, these grafts were made during embryonic stages and, in addition to the epithelial anlage of the pituitary, contained the floor of the diencephalon and the roof of the archenteron (Blount, '50). Thus hypothalamic tissue may have been present in the grafts. This experiment does not conclusively indicate, therefore, whether the *A. tigrinum* graft replaced a defective *A. mexicanum* pituitary, hypothalamus, or both. The data of the present experiment indicate that the defect is in the hypothalamus. When axolotl pituitaries were transplanted to newts, signs of thyroid activation were observed, thus indicating that the axolotl pituitary gland has the inherent ability to secrete TSH. It is likely that TSHRF from the newt's hypothalamus reaches the xenograft through the circulation. This is also apparently true in the case of the pituitary homograft (Dent, '66). Etkin and Sussman ('61) demonstrated that when a mechanical barrier is placed between the hypothalamus and pituitary of *Ambystoma maculatum* larvae, metamorphosis is prevented unless blood circulation is restored between the two tissues. Thus, TSH secretion by the *A. maculatum* pituitary apparently requires TSHRF. That TSHRF is necessary for TSH production from both ectopic and intact newt pituitary glands is suggested by the experiments of Tassava *et al.* ('68) who found that adult newt pituitary glands, grafted ectopically, induced precocious metamorphosis of *A. maculatum* and *A. laterale* larvae, two species which normally undergo metamorphosis, but not *A. mexicanum* larvae. This observation would be expected if indeed the hypothalamus of larval *A. mexicanum* did not produce TSHRF.

A combination of prolactin (0.015 U/2 days) and thyroxine (1×10^{-7} conc. in aquarium water), but neither alone, was found to significantly improve survival of hypophysectomized adult newts and also maintain the normal ability to regenerate amputated limbs (Tassava, '69). This fact, and ultrastructural evidence (Dent and Gupta, '67), suggest that the intact and the ectopic pituitary of the adult newt secrete both TSH and prolactin. Since xenografts of larval axolotl pituitaries are as effective as newt pituitary homografts in improving survival and limb regeneration of hypophysectomized newts, it may be that prolactin, in addition to TSH, is secreted by the larval axolotl pituitary when grafted in the newt. Whether a prolactin-like hormone is produced *in situ* by the larval axolotl pituitary, and what its function is, remain to be determined. There is evidence that pituitary hormones are not essential to survival or limb regeneration of larval axolotls (Tassava *et al.*, '68) and, furthermore, hypophysectomized axolotls appear to grow at a normal rate (Sprague and Tassava, unpublished observations) as is true of hypophysectomized frog tadpoles (Just and Kollros, '68). Ultrastructural studies of intact and transplanted larval axolotl pituitaries may help to clarify these problems (Connelly *et al.*, '68).

Limb regeneration of hypophysectomized newts with larval axolotl pituitary xenografts was normal in all respects. This finding is contrary to the report of Schotté ('61) that pituitary glands from larval newts or larval *Ambystoma* were unable to support limb regeneration. It may be that the condition of neoteny is related to the ability of the axolotl pituitary to maintain survival and limb regeneration of hypophysectomized newts. Gona ('67) and Derby and Etkin ('68) have shown that exogenous prolactin can inhibit metamorphosis of frog tadpoles by acting both at the level of the thyroid and the peripheral tissue. It is possible that the pituitary of the axolotl secretes more prolactin than the pituitary of other urodeles and thus desensitizes the peripheral tissues to thyroxine. However, axolotl tissues of almost any age readily respond to thyroxine which indicates little if any peripheral inhibition of thyroxine activity (Derby and Etkin, '68). Schotté and Droin ('65) reported that the pituitary of the recently metamorphosed newt (brown eft) was ineffective in promoting limb regeneration of hypophysectomized newts, whereas the pituitary of the more well established land phase of the newt (red eft) was effective. It would be of interest, therefore, to test whether pituitaries of younger axolotls would promote survival

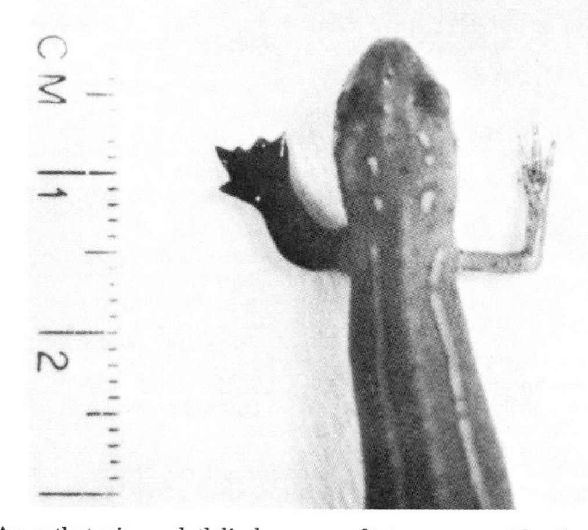

Fig. 9 An orthotopic axolotl limb xenograft to a newt. The fingers of this axolotl limb xenograft began regressing 16 days after transplantation until by 22 days, when good vascularization was observed, only 50% of the original length remained. The distal end of the stump was amputated at this time. No additional regression occurred. By 30 days after amputation a four-digit regenerate was completed. A second amputation was performed 185 days after the initial transplantation at which time the normal right limb of the newt was amputated. Both limbs regenerated; however, the axolotl limb xenograft regenerated at a more rapid rate than the newt limb but at a rate comparable to that of axolotls of this size (snout-tail tip length 11 cm.). This picture was taken 175 days after transplantation.

and limb regeneration of hypophysectomized newts. Whatever the reason it is clear that, in the present case, limb regeneration of adult newts took place in a normal fashion with *larval* pituitary tissue as the source of hormones.

Skin allografts to adult newts are rejected in less than 40 days (Cohen, '66) and it is puzzling, therefore, why pituitary homografts (Dent, '67) and xenografts will survive for over four times that long. It is possible that skin and pituitary tissue have considerable antigenic differences with a more active and earlier immunologic response induced by the former (see White and Hildemann, '68). However, if this is so, newt skin would seem to be unique in this property because axolotl skin sleeve grafts to newt limbs (denuded of newt skin) were found to survive for over 60 days and, in one case, an entire larval axolotl limb survived for 200 days grafted orthotopically on a newt. Upon amputation, this limb xenograft also underwent regeneration in a normal fashion (Tassava, unpublished observations, and fig. 9).

ACKNOWLEDGMENTS

This work was supported by NIH Training Grant HD 00135 while I was a postdoctoral trainee and by NSF Grant GB-7748 and NIH Grant MB-04128-07 administered by Dr. C. S. Thornton, Department of Zoology, Michigan State University. I wish to thank Dr. Thornton for helpful suggestions during this investigation and for critically reading the manuscript.

LITERATURE CITED

Blount, R. F. 1950 The effects of heteroplastic hypophyseal grafts upon the axolotl, *Ambystoma mexicanum*. J. Exp. Zool., *113:* 717-740.

Cohen, N. 1966 Tissue transplantation immunity in the newt, *Diemictylus viridescens*. II. The rejection phase: first and second-set allograft reactions and lack of sexual dimorphism. J. Exp. Zool., *163*(2): 173–190.

Connelly, T. G., R. A. Tassava, and C. S. Thornton 1968 Survival and limb regeneration of prolactin treated hypophysectomized newts. J. Morphology, *126*(3): 365–371.

Connelly, T. G., S. K. Aggarwal, and C. S. Thornton 1968 Fine structural differences between the normal and the ectopically transplanted axolotl pituitary. Abstracts of papers: 8th Annual Meeting of the American Society for Cell Biology.

Dent, J. N. 1966 Maintenance of thyroid function in newts with transplanted pituitary glands. Gen. Comp. Endo., *6:* 401–408.

——— 1967 Survival and function in hypophysial homografts in the spotted newt. Amer. Zool., *7*(4): 714.

Dent, J. N., and B. J. Gupta 1967 Ultrastructural observations on the developmental cytology of the pituitary gland in the spotted newt. Gen. Comp. Endo., *8:* 273–288.

Derby, A., and W. Etkin 1968 Thyroxine induced tail resorption *in vitro* as affected by anterior pituitary hormones. J. Exp. Zool., *169*(1): 1–8.

Geschwind, I. I. 1967 Molecular variation and possible lines of evolution of peptide and protein hormones. Amer. Zool., *7:* 89–108.

Gona, A. 1967 Prolactin as a goitrogenic agent in amphibia. Endo., *81:* 748–754.

Just, J. J., and J. J. Kollros 1968 Effects of hypophysectomy, thyroxine and pituitary hormones on growth of *Rana pipiens* larvae. Amer. Zool., *8*(4): 762.

Liversage, R. A. 1967 Hypophysectomy and forelimb regeneration in *Ambystoma opacum* larvae. J. Exp. Zool., *165:* 57–70.

Lynn, G. D., and H. E. Wachowski 1951 The thyroid gland and its functions in cold-blooded vertebrates. Quart. Rev. Biol., *26:* 123–168.

Pearse, A. G. E. 1961 *Histochemistry: Theoretical and Applied.* Little, Brown and Company, Boston.

Prahlad, K. V., and L. E. DeLanney 1965 A study of induced metamorphosis in the axolotl. J. Exp. Zool., *160:* 137–146.

Schotté, O. E. 1961 Systemic factors in initiation of regenerative processes in limbs of larval and adult amphibians. In: *Molecular and Cellular Structures,* 19th Growth Symposium. Editor, D. Rudnick. The Ronald Press Co., N. Y.

Schotté, O. E., and A. Droin 1965 The competence of pituitaries and limb regeneration during metamorphosis of *Triturus viridescens.* Rev. Suisse de Zool., *72:* 205–223.

Schotté, O. E., and A. Tallon 1960 The importance of autoplastically transplanted pituitaries for survival and for regeneration of adult *Triturus.* Experientia *16:* 72–76.

Stone, L. S. 1967 An investigation recording all salamanders which can and cannot regenerate a lens from the dorsal iris. J. Exp. Zool., *164:* 87–104.

Tassava, R. A. 1968 Limb regeneration of hypophysectomized adult newts. Amer. Zool., *8*(4): 785–786.

——— 1969 Hormonal and nutritional requirements for limb regeneration and survival of adult newts. J. Exp. Zool., *170* (1): 33–54.

Tassava, R. A., F. J. Chlapowski, and C. S. Thornton 1968 Limb regeneration in *Ambystoma* larvae during and after treatment with adult pituitary hormones. J. Exp. Zool., *167*(2): 157–163.

White, F., and W. H. Hildemann 1968 Allografts in genetically defined rats: difference in survival between kidney and skin. Science, *162* (3859): 1293–1295.

General and Selective Inhibition of Amphibian Regeneration by Vinblastine and Dactinomycin

R. T. Francœur

Introduction

Vinblastine (VLB), an alkaloid extract of *Vinca rosea* Linn. marketed under the trade name Velban, has been used extensively since 1961 in carcinostatic treatment of Hodgkin's disease and other lymphomas as well as both monocytic and erythro-leukemias. Antitumoral effects have also been reported for the drug in treating carcinomas of the lung, bronchus, breast, stomach, colon, rectum, cervix and uterus (Garattini and Sproston 1966; Horn and Hochman 1967).

Despite this extensive use and the similarities between carcinomatous neoplasms, embryonic and regenerating tissues pointed out by Nowinski (1960) and others, only one report has appeared on the inhibitory effects of VLB on regenerating tissue. Cutts, Beer and Noble (1960) reported no inhibition of regeneration in partially hepatectomized rats.

Dactinomycin, an antibiotic isolated by Vining and Waksman in 1954 from soil Streptomyces, has had a much more limited oncolytic use, primarily against carcinomas such as Mills tumor. Extensive studies of its strong growth inhibition with both vertebrate and invertebrate tissue, *in vivo* and *in vitro*, show the primary effect to be an inhibition of nucleic acid syntheses dependent on DNA.

In the tail regeneration of larval amphibians Dactinomycin greatly inhibits developmental (differentiation) processes. Unlike VLB which causes metaphase arrest, Dactinomycin has little inhibitory effect on mitosis (Wolsky and Wolsky, 1961, 1964). Yet the inhibition of regeneration is extensive, due to lack of differentiation (regeneration blastema formation), and the inhibition of regeneration follows a similar in both *Ambystoma* and *Rana pipiens* larvae (WOLSKY and VAN DOI, 1965).

The purpose of the present study is to compare the inhibitory effects of VLB and Dactinomycin on the regenerating tails of larval *Rana pipiens* and to examine possible selective action of VLB.

Materials and Methods

Design of the Experiment

This study is based on 4 fixed variables: the 2 drugs, VLB and Dactinomycin; 7 drug concentrations –20, 10, 5, 2.5, 1.25 and 0.67 μg/ml plus a control; the 2 age levels –5 and 10 day old larvae; and the 3 temperatures of 15, 20 and 25° C. A fifth variable, random or dependent, is the amount of tissue regenerated.

Research involved 4,200 larval *Rana pipiens* arranged in 12 experimental blocks, each containing 350 animals. Each block consisted of 50 animals (1 group for of the 6 drug concentrations) plus a spring water control group. Fifty 5 and fifty 10 day old animals were tested in each of the 7 concentrations at each of the 3 temperatures so that all possible combinations were covered.

Materials and Support. Rana pipiens males and injected gravid females were obtained commercially. Dactinomycin was supplied by Merck, Sharp and Dohme Research Laboratory in the crystaline form, Lot No. L554651–0–10; VLB by Eli Lilly Pharmaceuticals in the crystalline sulfate form, Lot No. 862543, potency 92.4% in the dry condition.

This research was supported by Air Force Office of Scientific Research Grant No. 997–66.

Procedures

Injected gravid *Rana pipiens* were ovulated and their eggs fertilized according to the method of RUGH (1952). After hatching, the larvae were maintained in temperature control chambers with a constant humidity and fluorescent lighting on a cycle of 13 h. of light and 11 h. of dark.

Five or 10 days after hatching, the distal halves of the tails of 700 larvae kept at one of the 3 temperatures were amputated. Twenty-five such animals were then placed in glass culture dishes with 250 ml of the desired drug solution (VBL or Dactinomycin) or spring water. Two culture dishes, 50 animals, were kept for 4 days in each of the 6 concentrations, in addition to 50 control animals in pure spring water for each drug series.

Original concentration was maintained by adding fresh spring water to compensate for evaporation. Animals dying during the 4 day treatment were discarded. A small piece of Elodea in each dish served as food.

At the end of the 4 day treatment all surviving animals were fixed in neutral-buffered 10% formalin. The whole animals were then carried through a modified haematoxylin-eosin staining, each group of 50 in an individual Lipshaw Autopsy Basket (No. 342). Immersion in Harrison's haematoxylin was limited to 2½ min instead of the 15 min customary for sectioned tissue. The shorter time required for proper staining of the whole animal is probably due to the fact that in the unsectioned tail we are dealing with many layers of cells including the rather compact central muscle. An exposure longer than 2½ min gives a cumulative nuclear stain so heavy the histological details are completely obscured. The rest of the staining technique was standard. After complete dehydration, 24 h in absolute alcohol, and Xylene clearing, the stained tails were amputated and mounted on slides.

The amount of tissue regenerated was ascertained by enlarging the tail whole mount 47× with a bioscope and measuring the projected image with a compensating polar planimeter, using the notches on the tail edge as markers for the line of amputation. Animals dying during the 4 day test were recorded as having regenerated no tissue. Of the animals that survived 69 failed to mount properly and had to be recorded as lost regeneration data. Computer calculations for mean regeneration and standard deviation for each group compensated for these lost data by using (computer) cells of various sizes.

Photomicrographs were made at 10× with an automatic exposure Zeiss Photomicroscope and Agfachrome 135 mm film.

Observations

I. General Inhibition by the two Drugs

Tables I and II summarize the mean regeneration, standard deviation and percentage of inhibition for each group of 50 animals treated with 20 and 0.67 μg/ml solutions of the 2 drugs.

Table I. Mean regeneration in sq.cm for the projected image of regenerated tails enlarged 47× and measured with a compensating polar planimeter. Data are given for animals subjected to the highest and lowest concentrations of the two drugs and for their control groups. An * indicates mean regeneration figure for animals treated with 10 μg/ml in a group where all animals subjected to 20 μg/ml died during the test. An ** indicates survival of only one of the 50 animals in a group.

				Vinblastine				Actinomycin D			
		Control		20 μg/ml		0.67 μg/ml		20 μg/ml		0.67 μg/ml	
Age	Temp.	Mean Regen.	St. Dev.	Mean Regen.	St. Dev.	Mean Regen.	St. Dev.	Mean Regen.	St. Dev.	Mean Regen.	St. Dev.
5 Day Old	15° C	11.1	3.7	6.5	2.4	10.0	2.0	3.2	1.6	10.6	3.3
	20° C	32.8	6.0	12.8	3.3	31.5	5.4	1.4	0.2	31.9	6.6
	25° C	21.1	6.7	7.8*	3.3	22.1	6.0	6.1**	0.0	20.3	7.2
10 Day Old	15° C	9.2	2.3	5.3	1.7	8.8	2.3	—	—	—	—
	15° C	11.4	2.1	—	—	—	—	1.9	0.4	9.3	1.9
	20° C	17.3	3.6	8.1	2.0	14.1	2.3	1.8	0.6	14.4	3.0
	25° C	15.7	5.7	6.7	2.2	15.5	5.9	2.0*	0.4	11.8	4.6

Table II. The percentage of general inhibition observed in animals treated with 20 and 0.67 μg/ml with the specific control groups as the basis for comparison.

Age	Temperature	VLB concentration 20	VLB concentration 0.67	Actinomycin concentration 20	Actinomycin concentration 0.67
5 Day Old	15° C	—42%	—11%	—71%	—5%
	20° C	—61	— 4	—96	—3
	25° C	—63	+ 4	—74	—4
10 Day Old	15° C	—42	— 4	—90	—18
	20° C	—53	—18	—90	—17
	25° C	—57	0	—87	—25

With both VLB and Dactinomycin, inhibition of regeneration varies directly with the drug concentration: the higher the concentration, the greater the inhibition.

Under comparative conditions Dactinomycin is generally a stronger inhibitor of regeneration than VLB except in the lower concentrations where the difference is not significant. At 20 and 10 μg/ml Dactinomycin is between 18 and 110% more effective in producing regeneration inhibition.

At 20 μg/ml the inhibitory effect of VLB increases with temperature in both age groups. The reverse is the case with 5 day old animals treated with 0.67 μg/ml VLB where a temperature increase brings a decrease in inhibition. No similar correlation is evident with Dactinomycin.

At the highest concentration VLB is slightly more inhibitory in the 5 day old group than with 10 day old animals. The reverse is the case with Dactinomycin.

II. Interactions

Taken individually, each of the four fixed variables, age, temperature, drug and drug concentration, significantly influence the inhibition of regeneration. These same factors interact significantly in combination except the 2-way interaction of age and drug and the 3-way interaction of age, drug and drug concentration. The level of significance used here is 0.001; calculations were made in a 4-way analysis of variance in a SDS 9300 binary digital computer according to the program outlined by Francœur and Wilber (1967).

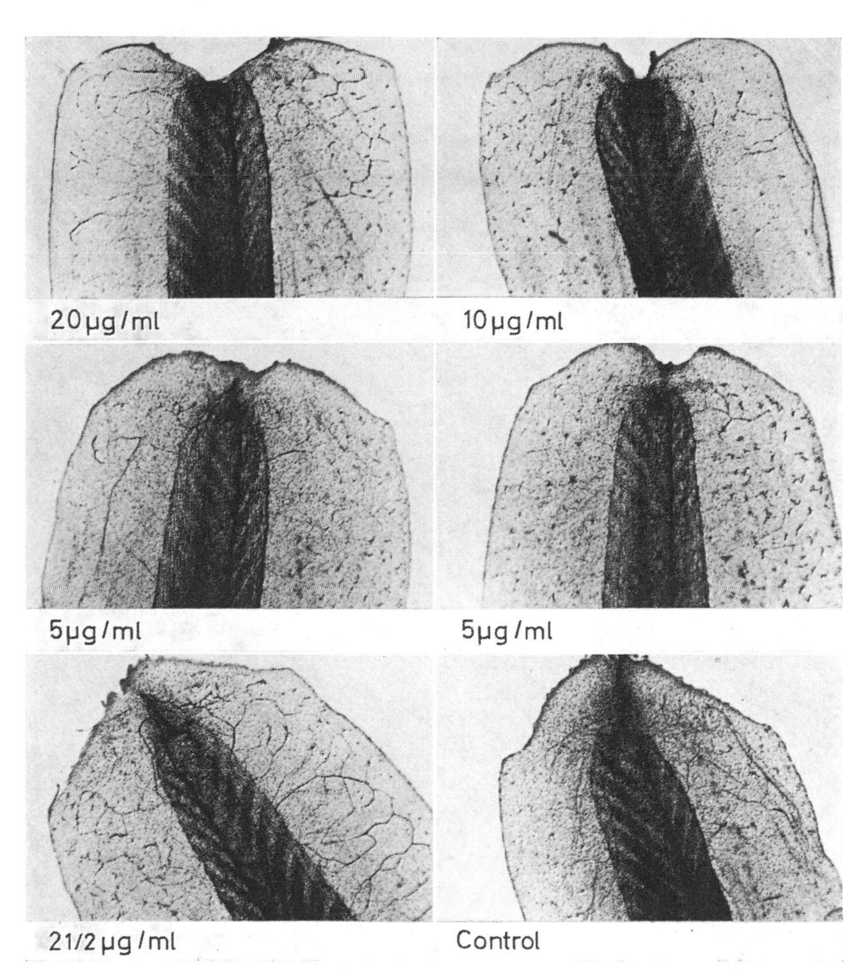

Fig. 1. Selective inhibition of muscle and notochordal regeneration in 10 day old larvae treated with VLB at 20° C. The differential effect decreases as the drug concentration is lowered.

III. Selective Inhibition of Regeneration by VLB

Unlike Dactinomycin which inhibits all regenerative processes, VLB exhibits a selective action, at least under certain circumstances in the present study.

With VLB treatment at 20° C general inhibition of tail regeneration is in proportion to the drug concentration as expected. However, in 10 day old animals regeneration of epithelial tissue is far less retarded by the drug than is regeneration of the central muscle and notochord

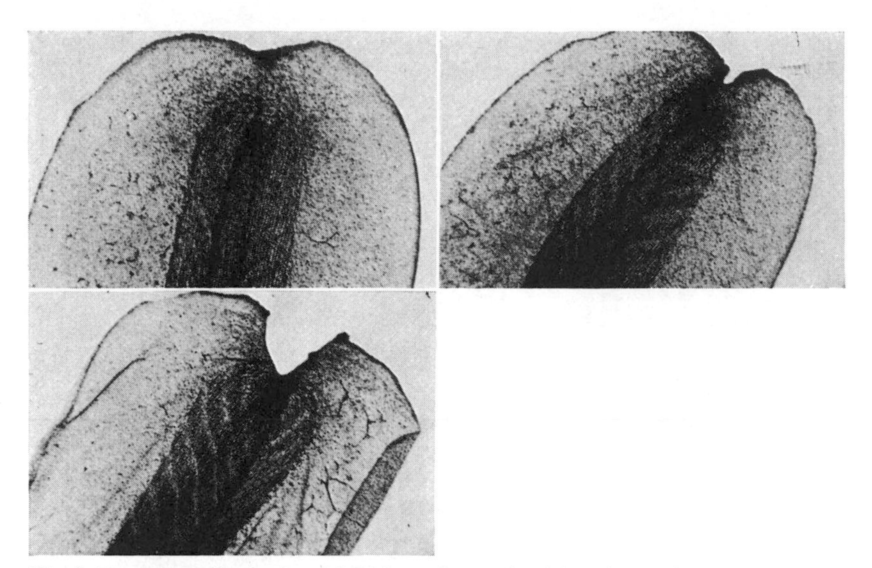

Fig. 2. Examples of selective inhibition of notochordal and muscle regeneration in 5 day old animals treated with 20 μg/ml VLB at 20° C. Animals treated with lesser concentrations of VLB do not show this selective effect.

which are inhibited almost completely. Most 10 day old larvae treated with 20 and 10 μg/ml VLB at 20° C show this effect, while about half those subjected to 5 μg/ml exhibit selective inhibition. Concentrations below 5 μg/ml VLB do not show the effect at all (Fig. 1).

Five day old animals treated at 20° C with VLB show a similar but much weaker response in this selective action. Only a few animals in the 20 μg/ml solution show differential inhibition (Fig. 2).

IV. Maximum Inhibiting Effect

Graphs relating drug concentration and the amount of tissue regenerated indicate a maximum inhibiting effect for Dactinomycin of approximately 10 μg/ml for 10 day old animals at all 3 temperatures and for 5 day old larvae treated at 15 and 20° C. Graphs for 5 day old larvae treated with Dactinomycin at 25° C and with VLB at 20 and 25° C suggest a similar maximum inhibitory dosage of 10 μg/ml. Beyond this concentration inhibition of regeneration does not increase significantly. The main effect of higher concentrations is greater toxicity and mortality. Other graphs, for VLB, are inconclusive in this respect (Tab. III).

Table III. Summary of graphs relating the amount of regeneration and drug concentrations for each of the 12 experimental blocks. Those graphs clearly indicating a maximum effective concentration of 10 μg/ml are marked +; those giving a questionable indication, ?, and those graphs which are inconclusive in this aspect, —.

		VLB	Dact.
5 Day Old	15° C	—	+
	20° C	?	+
	25° C	?	?
10 Day Old	15° C	—	+
	20° C	—	+
	25° C	—	+

Discussion

I. General Growth Inhibition

Biochemical studies have thus far failed to uncover the mechanisms behind the oncolytic effects of VLB. Similar studies have also failed to disclose the mechanisms that lead to metaphase arrest and the accumulation of cells in the classic C-mitotic pattern, the most obvious cytological effect of VLB (Neuss *et al.*, 1964; Sentein, 1964; Svoboda, 1966). Neuss *et al.* (1964) have shown however that whatever the mechanisms of VLB may be they are not directly related to action on the nucleic acids.

Dactinomycin, on the other hand, is known to complex directly with guanine in the minor groove of the DNA helix, thus inhibiting the synthesis of nucleic acids, particularly RNA, dependent on DNA (Hurwitz *et al.*, 1962; Reich, 1964).

The present experiment shows that Dactinomycin is 18 to 110% more effective than VLB in general inhibition of regeneration. This may be due to Dactinomycin's direct action on the nucleic acids.

In this context a comparison of molecular weights is interesting since the two drugs have approximately the same maximum effective dose and their mode of administration in the present experiment is the same, passive diffusion. Equal volumes of the drugs contain 4 molecules of VLB for every 3 molecules of Dactinomycin. One might suspect then that VLB would be $^{1}/_{3}$ more effective in inhibiting regeneration. If passive diffusion is the pathway of absorption we would also expect the smaller VLB molecules to enter the regenerating tail more rapidly, thus adding to its effectiveness. Yet the opposite is the case: Dactinomycin is much more effective.

This would suggest that molecular size and passive diffusion are not as important in the inhibitory effect of the two drugs as other factors such as iso-electric point, lipid solubility, molecular charge, active transport mechanisms and especially the site of activity. A single Dactinomycin molecule, acting directly on the DNA complex, is much more effective in broad inhibition of growth and regeneration than VLB which seems to act on more diffuse or secondary processes of enzyme and/or protein activity or synthesis. The same differential effect would prevail if VLB affects membrane activity.

II. Maximum Inhibiting Dose

Five of the 12 graphs relating drug concentration and regeneration give strong evidence of a maximum inhibiting dose of 10 μg/ml in a response curve approximating a rectangular hyperbola. Three other graphs suggest a similar curve. This corroborates WILBER's suggestion (1965) that 'direct experimental data support the theory favoring the (rectangular) hyperbola as a widely applicable generalization describing the effects of pharmacologically active agents on living protoplasm.'

III. Selective Inhibition

Whereas Dactinomycin inhibits all regenerative processes and results in much smaller animals after the 4 day treatment, VLB treated larvae are close to the control in body size and show a differential or selective pattern of regeneration inhibition. Epithelial tissue is least affected by VLB while notochord and muscle regeneration are selectively inhibited. This would indicate that VLB in some way affects the differentiation of tissue derived from mesoderm and does not inhibit ectodermal tissue. This is in agreement with HODES' (1960) report of a selective toxicity for VLB which affects bone marrow and other tissues of mesodermal origin but not intestinal tissues (which are a combination of mesodermal and endodermal derivatives), and with a report of VLB's selective inhibition of blood islands in regenerating amphibian tails (FRANCŒUR and WILBER, 1968).

References

CUTTS, J.H., BEER C.T. and NOBLE R.L.: Biological properties of Vincaleukoblastine, an alkaloid in *Vinca rosea* Linn. with reference to its antitumor action. Cancer Res. *20:* 1023–1031 (1960).

FRANCŒUR, R.T. and WILBER C.G.: Comparative effects of Vinblastine and Dactinomycin on regenerating larval *Rana pipiens* tails. DA 4–193MD–2627. U.S. Army Med. Res. and Development Command. Defense Documentation Center, Cameron Station, Va. (1967).

FRANCŒUR, R.T. and WILBER, C.G.: Amphibian regeneration and the teratogenic effects of Vinblastine. Oncology *22* (in press).

GARATTINI, S. and SPROSTON, E.M.: Antitumoral effects of Vinca alkaloids, Int. Congr. Ser. No. 106. (Excerpta Medica Foundation, Amsterdam 1966).

HODES, M.E., ROHN R.J. and BOND, W.H.: Vincaleukoblastine I. Preliminary clinical studies. Cancer Res. *20:* 1041–1049 (1960).

HORN, Y. and HOCHMAN, A.: The alkaloids of *Vinca Rosea* Linn. in malignant tumors. Oncology *21:* 214–220 (1967).

HURWITZ, J., FURTH, J.J., MALAMY, M. and ALEXANDER, M.: The role of deoxyribonucleic acid in nucleic acid synthesis III. Inhibition of enzymatic synthesis of ribonucleic acid and deoxyribonucleic acid by Actinomycin D and proflavin. Proc. nat. Acad. Sci. *48:* 1222–1230 (1962).

NEUSS, N., JOHNSON, I.S, ARMSTRONG, J.G. and JANSEN, C.J.: The *Vinca* Alkaloids. Adv. chemother. Vol. 1 (Academic Press, New York 1964).

NOWINSKI, W.W.: Fundamental aspects of normal and malignant growth (Elsevier, New York 1960).

REICH, E.: Actinomycin: Correlation of structure and function of its complexes with purines and DNA. Science *143:* 684–689 (1964).

RUGH, R.: Experimental embryology (Burgess, Minneapolis 1952).

SENTEIN, P.: L'action de la vincaleukoblastine sur la mitose chez *Triturus helveticus* Raz. Chromosoma *15:* 416–456 (1964).

SVOBODA, G.H.: The current status of research on the alkaloids of *Vinca rosea* Linn. (*Catharanthus roseus* G. Don). In: Garattini and Sproston (1960).

WILBER, C.G.: Biology of water toxicants in sublethal concentrations. U.S. Dept. Health, Ed. and Welfare. 3rd Seminar on Biol. Problems in Water Pollution, 1965.

WOLSKY, A. and VAN DOI, N.: Effect of Actinomycin on regeneration processes in amphibians. Trans. N.Y. Acad. Sci. *27:* 882–893 (1965).

WOLSKY, A. and WOLSKY, M.I.: The effect of Actinomycin D on the development of Arbacia eggs. Biol. Bull. *121:* 414 (1961).

WOLSKY, A. and WOLSKY, M.I.: An analysis of the actinomycin induced cleavage delay in sea urchin eggs. Z. Zellforsch. *63:* 960–964 (1964).

Amphibian Regeneration and the Teratogenic Effects of Vinblastine[1]

R. T. Francœur and Ch. G. Wilber

Introduction

Vinblastine (VLB), an alkaloid extract of *Vinca rosea* Linn. marketed by Eli Lilly Pharmaceuticals under the trade name Velban, has been used extensively since 1961, in carcinostatic chemotherapy of Hodgkin's disease and other lymphomas, and of both monocytic and erythro-leukemias. Antitumoral activity has also been reported for the drug in treating carcinomas of the lung, cervix, uterus, rectum, colon, stomach, breast and bronchus (Svoboda, 1966; Horn and Hochman, 1967).

Teratogenic effects of VLB have been observed in the embryonic development of two mammals. Anomalies observed in hamsters include embryonic mortality, microphthalmia, anophthalmia, spina bifidia, and skeletal defects, primary rib fusions and vertebral arch deformities (Ferm, 1963). In different strains of mice the teratogenic effects of VLB varied, including fetal death, exencephaly, polydactyly and harelip (Ohzu and Shoji 1965). Yet infants born to women treated for Hodgkin's disease with VLB during pregnancy show no anomalies (Armstrong *et al.*, 1964; Lacher, 1964; Rosenzweig *et al.*, 1964). Despite the similarities of carcinomatous neoplasms and embryonic and regenerating tissues (Nowinski, 1960), only one study has been made of the effects of VLB on regenerating tissues, showing no inhibition of regeneration in partially hepatectomized rats (Cutts *et al.*, 1960).

[1] This paper is dedicated as a modest tribute to Professor *Theodosius Dobzhansky*, of Rockefeller University, on the forthcoming occasion of his seventieth birthday. We hope his dedication to science and its broad humanistic impact will continue to influence young scientists everywhere for many years to come.

The purpose of the present study is to ascertain possible teratogenic effects in the tail regeneration of larval *Rana pipiens* treated with VLB.

Materials and Methods

Design of the Experiment

This study is based on 3 fixed variables: the seven drug concentrations –20, 10, 5, 2.5, 1.25 and 0.67 μg/ml plus a control; the 2 age levels –5 and 10 day old larvae; and the 3 temperatures of 15, 20 and 25° C.

Research involved 2100 larval *Rana pipiens* arranged in 6 experimental blocks, each containing 350 animals. Each block consisted of 6 groups of 50 animals each (1 group for each of the 6 concentrations ranging from 20 to 0.67 μg/ml) plus a spring water control group. Fifty 5 and fifty 10 day old animals were tested in each of the 7 concentrations at each of the 3 temperatures so that all possible combinations were covered.

Materials and Support

Rana pipiens males and injected gravid females were obtained commercially. The VLB was supplied by Eli Lilly Pharmaceuticals in the crystaline sulfate form, Lot 862543, with a potency of 92.4% in that condition.

This research was supported by Air Force Office of Scientific Research Grant No. 997–66.

Procedures

Injected gravid *Rana pipiens* were ovulated and their eggs fertilized according to the method of Rugh (1952). After hatching, the larvae were maintained in temperature control chambers with constant humidity and fluorescent lighting on a 24 h cycle of 13 h of light and 11 h of dark.

Five or 10 days after hatching, the distal halves of the tails of 350 larvae kept at one of the 3 temperatures were amputated. Twenty-five such animals were then placed in glass culture dishes with 250 ml of the desired VLB solution or spring water. Two culture dishes, 50 animals, were kept for 4 days in each of the 6 concentrations, in addition to 50 control animals kept in pure spring water.

The original concentrations were maintained by adding fresh spring water to compensate for evaporation. Animals dying during the 4 day treatment were removed immediately. A small piece of Elodea served as food in each culture dish.

At the end of the 4 day treatment all surviving animals were fixed in neutral-buffered 10% formalin. The intact animals were then carried through a modified haematoxylin-eosin staining in individual Lipshaw Autopsy Baskets (No. 342), one group per basket. Immersion in Harrison's haematoxylin was limited to 2½ min instead of the 15 min customary for sectioned tissue. The shorter time required for proper staining of the intact larvae is probably due to the fact that in the unsectioned tail we are dealing with many layers of cells including the rather compact central muscle. An exposure longer than 2½ min gives a cumulative nuclear stain so heavy that histological details are completely obscured. The rest of the staining technique was standard. After complete dehydration (24 h in absolute alcohol) and clearing in Xylene, the stained tails were amputated and mounted.

Photomicrographs were made at magnifications between 10× and 102×. Agfachrome 135 mm film was used with an automatic exposure Zeiss Photomicroscope. Higher magnifications (64 and 102×) were made with phase contrast lens.

Observations

I. *Anomalous blood vessel formation* is observed in animals treated with VLB at both 15 and 20° C, but not in animals maintained at 25° C.

Normally, blood vessels in an embryonic structure such as a larval *Rana pipiens* tail are formed in the peripheral regions by the anastomosis of many small blood islands. These dense, compact clumps of cells, with very little cytoplasm evident, gradually fuse and elongate into a fine vascular network. As the embryonic vessels elongate and become vascular, cells on the interior flake off to form erythrocytes. As vascular network formation proceeds, nuclei in the cells forming the vessel walls spread further and further apart so that the main portion of the wall is composed of a thin cytoplasmic layer connecting widely separated nuclei.

With VLB treatment at 20° C, blood vessel formation diverges from this normal pattern. The anomalous development increases with an increase in drug concentration and is more pronounced in 5 day old animals than it is in the 10 day old group.

In animals treated for 4 days with the higher concentrations of VLB at 20° C., the blood islands in the regenerating tail are much larger than normal and show little tendency to anastomosis or vascular network formation. These enlarged blood islands are much denser and more compact than those in the control animals. Nuclei are very tightly packed together with no cytoplasm evident. Blood vessel formation becomes more normal as the concentration of VLB decreases (Fig. 1).

At 15° C, animals treated with VLB for 4 days show a similar anomalous blood vessel formation with 2 interesting variations.

In 5 day old animals treated with VLB at 15° C, the effect on blood islands is again in direct proportion to the drug concentration with the higher concentrations of 20 and 10 μg/ml producing larger and more compacted islands. However, the lowest concentration used, 0.67 μg/ml, produces larger islands than either of the 2 highest concentrations (Fig. 2).

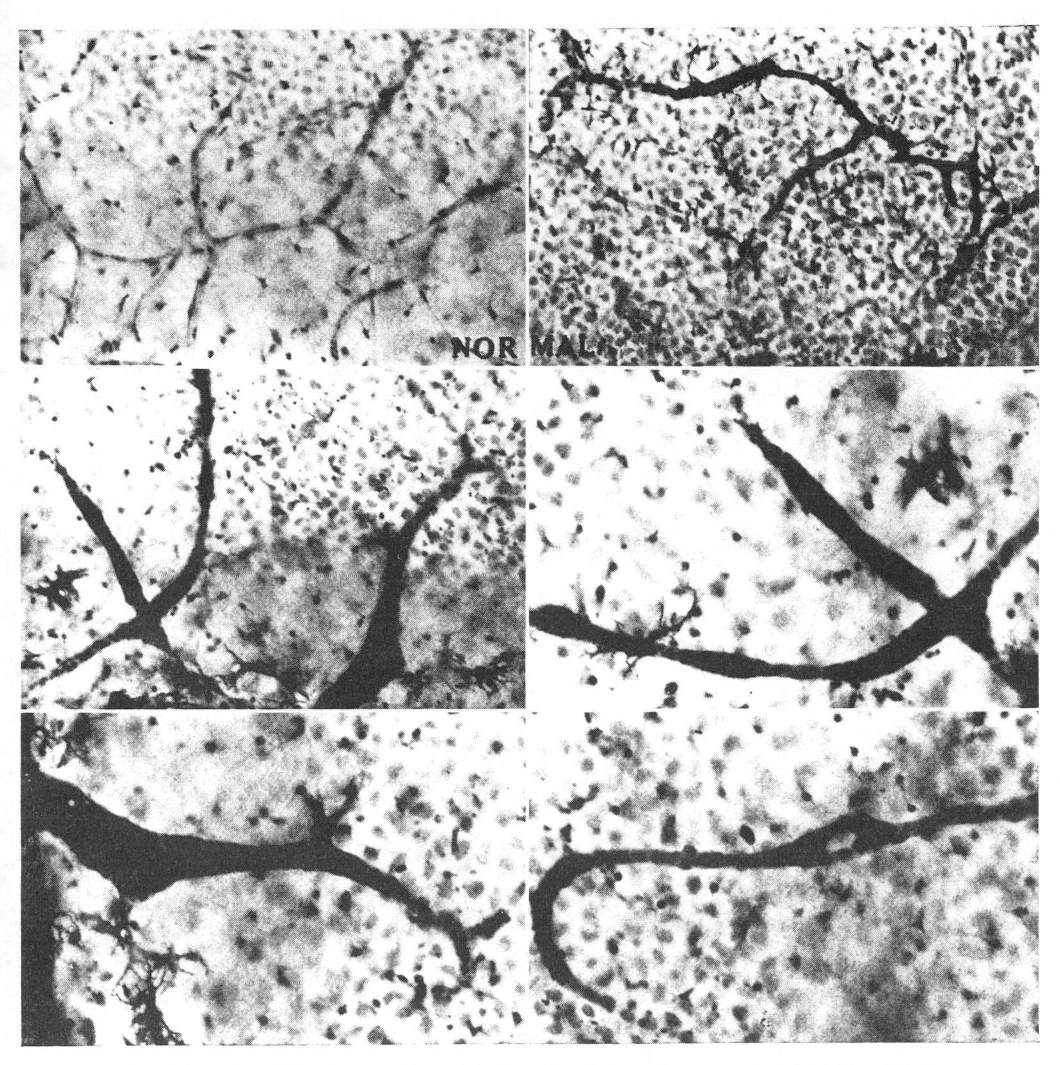

Fig. 1. Details of blood vessel formation in 5 day old tadpoles treated with VLB at 20° C. The control animals, 1 and 2, show a normal fine vascular network of blood vessels in the tail with anastomosis of heavy blood islands. The other photomicrographs show the greatly enlarged blood islands caused by VLB at a concentration of 10 μg/ml. No animals survived in the 20 μg/ml solution in this block.

In 10 day old animals treated at 15° C, the 2.5 μg/ml solution produces a vascular network approximating the normal pattern in every respect except the blood vessel diameter which is twice the size of vessels in the controls (Fig. 3).

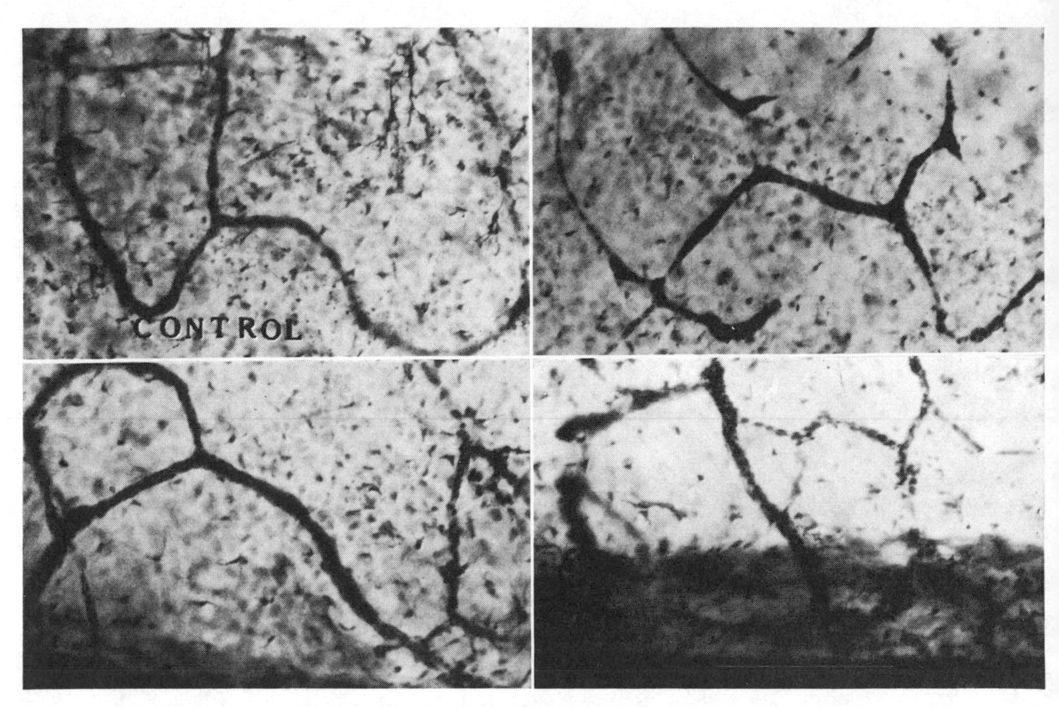

Fig. 2. Blood vessel details in 5 day old animals treated with VLB at 15° C. The anomaly is in direct proportion to the drug concentration except with the 0.67 μg/ml solution which produces an effect equal to or slightly more pronounced than that given by the highest concentration in this block. Microphotograph 1, Control; No. 2, VLB 20 μg/ml; Nos. 3 and 4, 0.67 μg/ml.

Blood vessels in both age groups treated at 15° C show a more normal tendency to anastamosis and vascular network formation than either age level treated at 20° C.

II. *Anomalies within the central tail muscle* are observed in approximately $^1/_3$ of the 1,800 animals treated with VLB. The striated muscle fibers just anterior to the point of amputation seem to separate from each other and spread apart. This effect is varied in degree, with some separations being only slight, others more pronounced and, in one animal, extreme (Fig. 4). The extreme example also shows an external fibrous growth, continuous with the central muscle bundle within the tail. The muscle fibers within the tail muscle of this animal are widely separated and apparently more branched than in the control. The external growth appears to be composed mainly of fibrous,

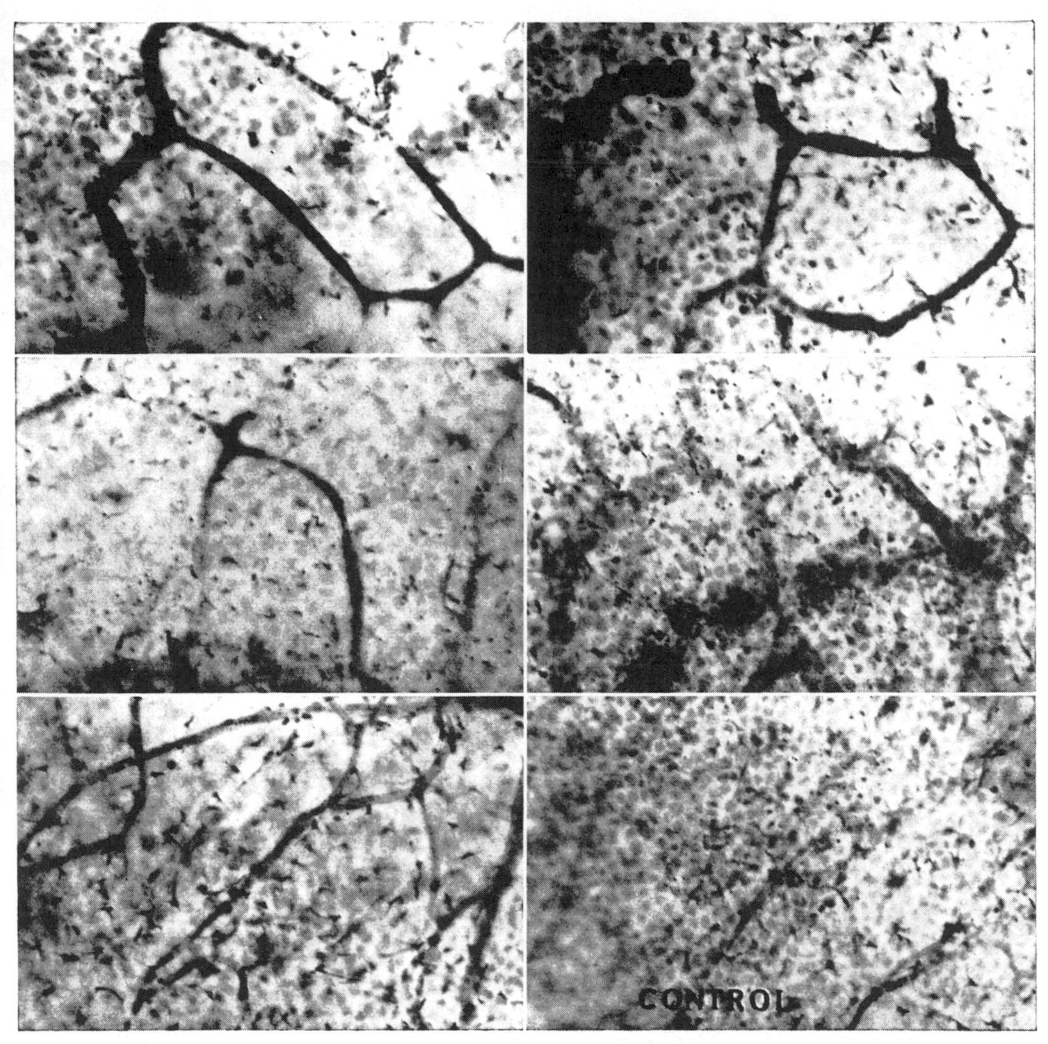

Fig. 3. Blood vessel details for 10 day old animals treated with VLB at 15° C. The anomalous blood vessel formation is generally similar to that observed in the two preceeding figures, except that at 2.5 μg/ml the blood vessels are abnormally dilated and approximately twice the diameter of blood vessels in the Control. Nuclei in the blood vessels of this exception are widely separated as in the Control and thus unlike the densely packed nuclei found at higher concentrations. Microphotograph No. 1, VLB 20 μg/ml; No. 2, VLB 10 μg/ml; No. 3, 5 μg/ml; No. 4, 2.5 μg/ml; No. 5, 1.25 μg/ml; No. 6, Control.

highly disoriented striated muscle tissue bound together by some connective tissue. Striations within the expanded muscle bundle are weaker in direct proportion to the fiber expansion (Fig. 5).

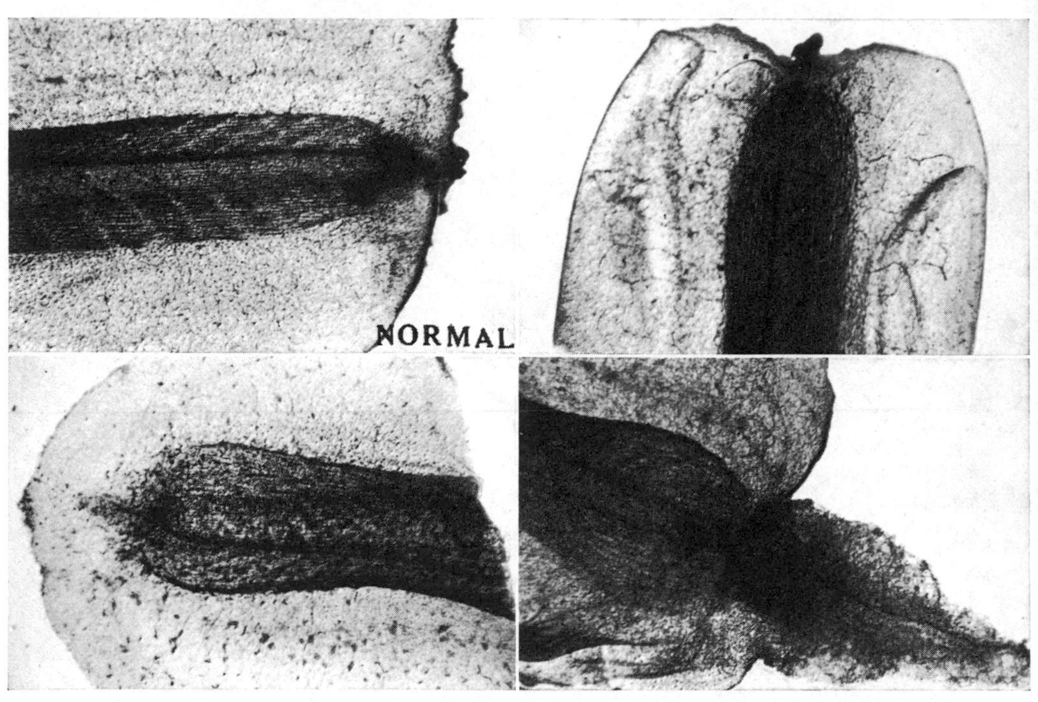

Fig. 4. Effect of VLB on the striated muscles of the regenerating tail. Separation of the fibers with some increase in branching and loss of striation pattern are evident in this and following figure. No correlation with temperature, age or drug concentration is apparent.

Discussion

The most obvious effect of VLB, both *in vivo* and *in vitro*, is the accumulation of metaphase-arrested cells in the classical C-mitotic pattern (SENTEIN, 1964). But according to SVOBODA (1966), 'it is highly unlikely that these cytological effects are responsible for the therapeutic (oncolytic) activity. The mechanism by which these alkaloids (VLB and the closely related Vincristine, Vinleurosine and Vinrosidine) inhibit tumor growth still remains unknown.'

In vivo and *in vitro* studies have shown neither inhibition nor stimulation of respiration or glycolysis, of protein or nucleic acid synthesis. In summarizing these studies, SVOBODA suggests that they are dealing with 'the biochemistry of mitotic arrest and not the oncolytic mechanism' (1966).

VLB is known to have a selective toxicity, affecting bone marrow and other tissues but not intestinal tissues (HODES *et al.*, 1960) and

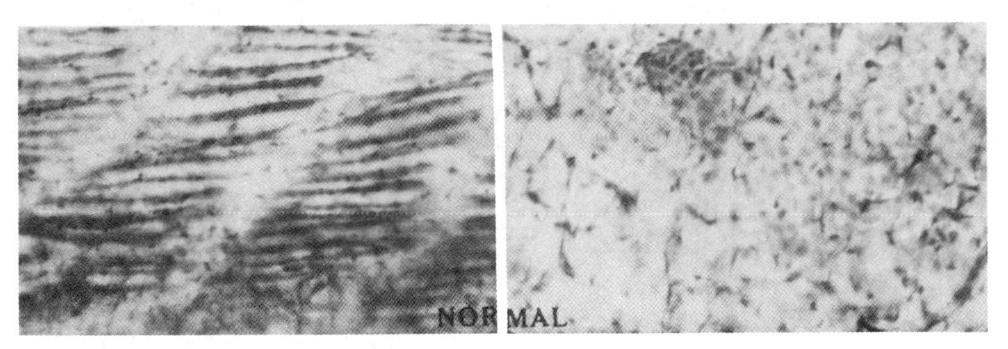

Muscle Fibers

Blood Vessels

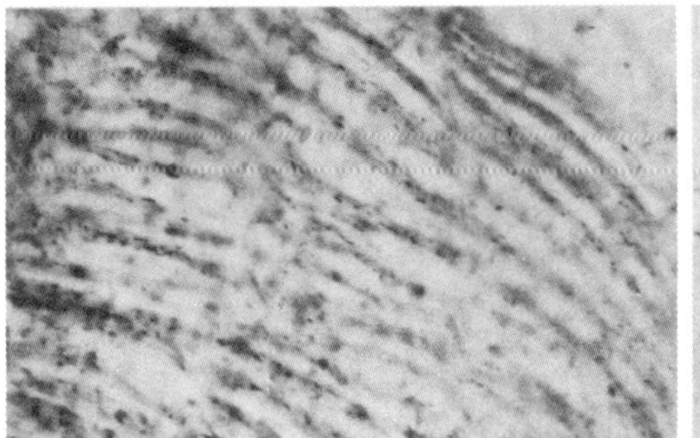

Separated Muscle Fibers inside the Extreme Tail

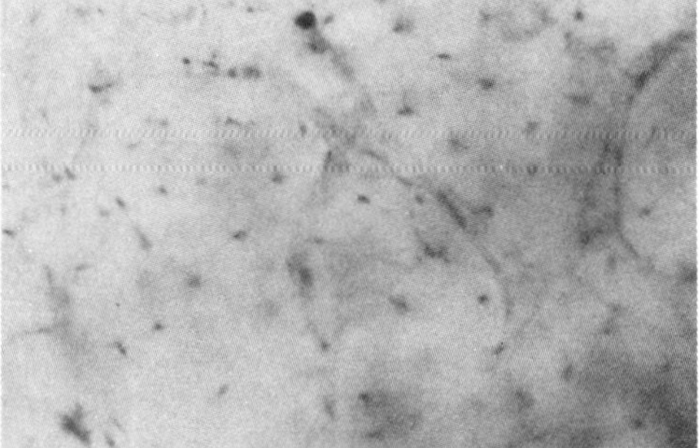

Dilated Blood Vessels in Extreme Example

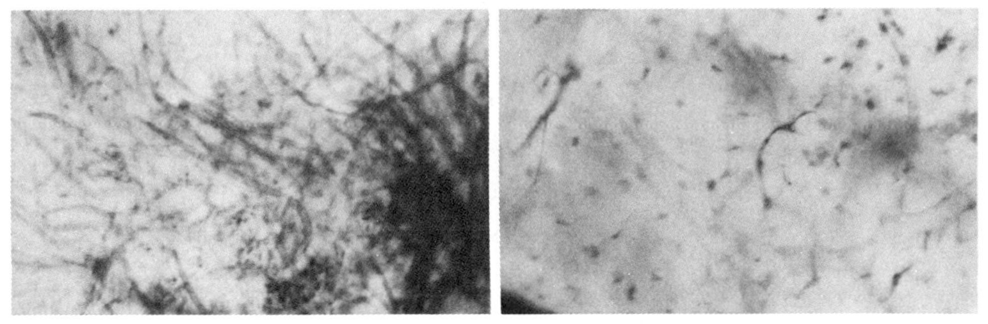

External Fibrous Growth

Dilated Blood Vessels in Extreme Example

Fig. 5. Microphotographs of muscles fibers and blood vessels in the tail of animal with anomalous external growth and the more normal development found in another animal in the same experimental group. All examples are from 10 day old larvae treated with 1.25 μg/ml VLB at 15° C. Separation of muscle fibers, loss of striation pattern and some branching are evident in the anomalous example along with dilation of the blood vessels.

inhibiting mesodermal and notochordal regeneration but not epithelial tissue (FRANCŒUR and WILBER, 1967). The present study shows that VLB fosters an accumulation of partially differentiated cells in the enlarged blood islands but prevents their further differentiation into vascular blood vessels.

Research by GARB and co-workers at the Albany (New York) Medical College have related tumor growth to a 'secretion by cancer cells of a chemical capable of inducing formation of new blood vessels by the host.' GARB quotes laboratory reports dating back more than 20 years to the effect 'that the rapid growth of tumor transplants is dependent upon the development of a rich vascular supply and... that an outstanding characteristic of the tumor cell is its capacity to elicit continuously the growth of new capillary endothelium from the host' (GARB, 1968).

In light of Garb's comments, the unknown antitumoral mechanism of VLB and the inhibition of blood vessel formation reported here, we would postulate that the oncolytic effects of VLB may be directly, and possibly primarily, related to its inhibition of *de novo* blood vessel formation on which the growth of the tumor depends. 'The circulatory pattern of neoplasms appears to be a fertile field for investigation. There is no small possibility that more knowledge in this sphere could be applied therapeutically, since checking this acquisition of a blood supply might restrict tumor growth' (Quoted by GARB, 1968).

Regeneration and Tumor Growth. The question of the relationship between regeneration and tumor formation has been repeatedly discussed (SEILERN-ASPANG, 1965; PIZZARELLO and WOLSKY, 1966). Two complementary statements sum up the relationship. First, tumor formation derives from an impaired or incomplete regenerative process, and second, regenerative processes may bring under control the autonomous growth of tumors.

Under normal conditions, tumors are found only in animals with low or no regenerative powers. Standard carcinogens applied to a regenerating animal do not cause tumors (PIZZARELLO and WOLSKY, 1966).

Furthermore, SEILERN-ASPANG (1965) has demonstrated the directive influence of mesenchymal tissue in the regenerative process, particularly in controlling the proliferation of epithelial tissue. In amphibian regeneration VLB inhibits regeneration of mesenchymal elements while allowing ectodermal components to continue proliferation (FRANCŒUR and WILBER, 1967).

The anomalous tuft of fibrous tissue observed in the extreme example of muscle fiber separation appears to be mesenchymal in derivation. Unlike other regenerating tails in the same experimental group, epithelial regeneration in the tufted tail is minimal, while mesenchymal elements apparently went uncontrolled. The fact that the muscle fiber separation can not be correlated with drug concentration, age or temperature and the occurrence of a tumorous neoplasm in a regenerating animal being treated with an antitumoral agent makes it difficult to suggest any explanation at present for this effect.

References

Cutts, J.H., Beer C.T. and Noble R.L.: Biological properties of Vincaleukoblastine, an alkaloid in *Vinca rosea* Linn. with reference to its antitumor action. Cancer Res. *20:* 1023–1031 (1960).

Ferm, V.H.: Congenital malformation in hamster embryos after treatment with vinblastine and vincristine. Science *141:* 426 (1963).

Francœur, R.T. and Wilber, C.G.: Comparative effects of Vinblastine and Dactinomycin on regenerating larval *Rana pipiens* tails. DA 4–193MD–2627. U.S.Army Med. Res. and Development Command. Defense Documentation Center, Cameron Station, Va., (1967).

Garb, S. Neglected approaches to cancer. Saturd. Rev. *51:* 22: 54–58 (1968).

Hodes. M.E., Rohn R.J. and Bond, W.H.: Vincaleukoblastine I. Preliminary clinical studies. Cancer Res. *20:* 1041–1049 (1960).

Horn, Y. and Hochman A.: The alkaloids of *Vinca Rosea* Linn. in malignant tumors. Oncology *21:* 214–220 (1967).

Lacher, M. Jr.: Vinblastine. Lancet *7347:* 1390 (1964).

Nowinski, W.W.: Fundamental aspects of normal and malignant growth (Elsevier, New York, 1960).

Ohzu, E. and Shoji, R.: Preliminary notes on abnormalities induced by Velban (vinblastine sulfate) in developing mouse embryos. Proc. jap. Acad. *41:* 321–325 (1965).

Pizzarello, D. J. and Wolsky, A.: Carcinogenesis and regeneration in newts. Experientia. *22:* 387–388 (1966).

Rosenzweig, A.I., Crews, Q. E. Jr. and Hopwood, H.G.: Vinblastine sulfate in Hodgkin's disease in pregnancy. Ann. intern. Med. *61:* 108–112 (1964).

Rugh, R.: Experimental embryology: a manual of techniques and procedures (Burgess, Minneapolis, 1952).

Seilern-Aspang, Y.: In: V.Kiortsis and H.A.L.Trampusch, Regeneration in animals and related problems (North Holland, Amsterdam, 1965).

Sentein, P.: L'action de la vincaleukoblastine sur la mitose chez *Triturus helveticus* Raz. Chromosoma *15:* 416–456 (1964).

Svoboda, G.H.: The current status of research on the alkaloids of *Vinca rosea* Linn. (*Catharanthus roseus* G.Don). In: S. Garattini and E.M. Sproston, Antitumoral effects of *Vinca rosea* alkaloids. (Intern. Congr. Ser. No. 106). (Excerpta Medica Foundation, Amsterdam, 1966).

Authors' addresses: R.T.Francœur, Ph. D., Biology Department, Fairleigh Dickinson University, *Florham Park-Madison, N.J. 07940* (USA) and Ch. G. Wilber, Ph. D., Colorado State University, Department of Zoology, *Fort Collins, Col.* (USA).

Regeneration of the Optic Lens In Amphibians

In Vitro Development of the Regenerating Lens[1]

Sara Eisenberg-Zalik and Violet Scott

INTRODUCTION

In the newt the pigment cells of the dorsal iris have the unique capacity of regenerating a new lens after the original has been surgically removed. Many aspects of lens regeneration have been studied *in vivo*, and these were reviewed recently by Stone (1960), Reyer (1954, 1962), and Yamada (1966, 1967a,b). After lentectomy the cells of the dorsal iris extrude their pigment granules (Eguchi, 1963, 1964; Karasaki 1964), synthesize RNA, protein, and DNA (Yamada and Karasaki, 1963; Yamada and Takata, 1963; Eisenberg and Yamada, 1966), go through a number of cell divisions (Eisenberg-Zalik and Yamada, 1967), and finally elongate and synthesize lens-specific proteins (Takata *et al.*, 1964, 1966).

Previously we reported *in vitro* culture of the dorsal iris undergoing lens regeneration (Eisenberg-Zalik and Meza, 1968). This paper extends our preliminary study and deals with the developmental capacities of the dorsal iris placed in culture at various stages of lens regeneration. Under the culture conditions used, it was found that once the lens placode forms (stage IV onward), some cells have the capacity to develop into lens fibers.

MATERIAL AND METHODS

Adult newts, *Triturus viridescens*, were lentectomized as described by by Eisenberg and Yamada (1966). The lens was removed from both eyes, and at the desired regeneration time the regenerate of one eye was placed in culture while the other was fixed and used as a control.

Dissection of the regenerates. One day prior to the experiments animals were placed in chlorine free water containing a penicillin-streptomycin mixture at a concentration of 200 units/ml (Microbiological Associates, Bethesda, Maryland). Newts were anesthetized in MS-222 (Sandoz), and one eye was fixed in Bouin's fixative while the other was used as the

[1] This work was supported by a grant from the National Research Council of Canada.

source of the regenerate. Lens regenerates were removed under aseptic conditions. The corneal incision which had been performed for lentectomy purposes was reopened with a sharp scalpel blade (Fig. 1, step 5), and extended to the lateral and median extremes of the eye (Fig. 1, step 6). With the aid of iris scissors this incision was continued upward and circumferentially along the upper margin of the eye until it joined the

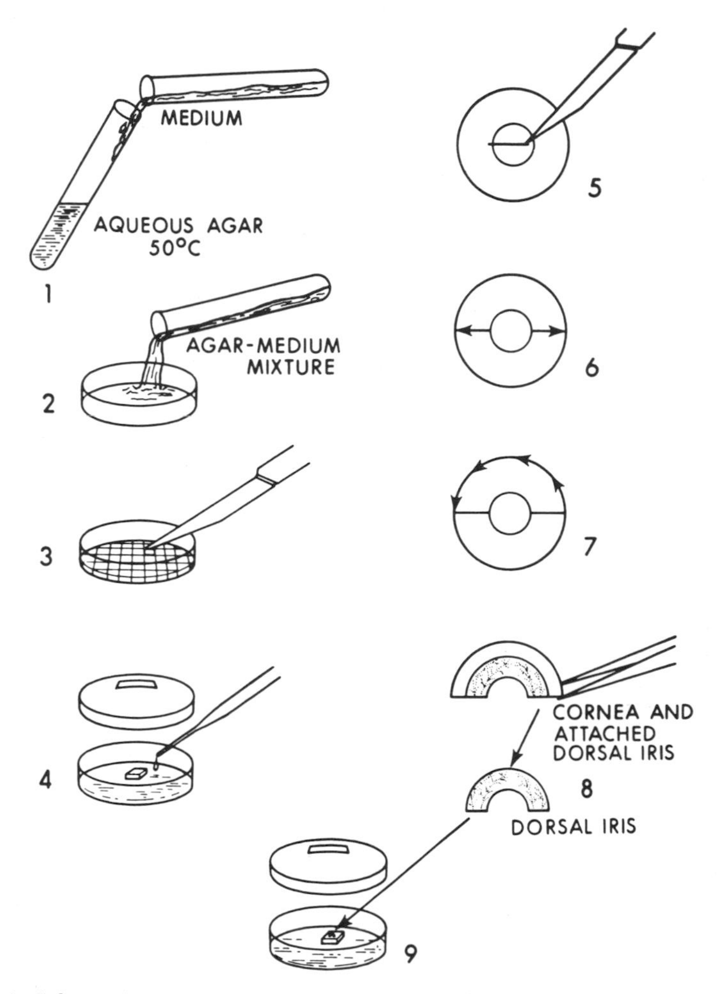

FIG. 1. Schematic representation of culture technique and dissection procedure. Steps 1–4 and 9, culture methods; 5–8, dissection of the regenerate; arrows indicate direction of the incision. For explanation see text.

initial corneal incision (Fig. 1, step 7). The cornea with the attached dorsal iris was removed and placed in culture medium, where the dorsal iris was gently separated from the corneal tissue with the aid of two scalpel blades (Fig. 1, step 8). Care was taken to obtain the dorsal iris free of other ocular tissues; it was then placed on agar blocks and cultured (Fig. 1, step 9).

Culture methods. Agar blocks were prepared by mixing a 4% aqueous agar solution and culture medium in a proportion 1:1 (v/v) (Fig. 1, step 1). The culture medium used was medium 199 with 15% fetal bovine serum (Microbiological Associates, Bethesda, Maryland). Penicillin and streptomycin were incorporated into the medium each at a concentration of 100 units/ml. The agar blocks were prepared and transferred to a sterile petri dish, which was then filled with a small amount of culture medium (Fig. 1, step 4). A detailed description of this method has been given previously (Eisenberg-Zalik and Meza, 1968). Tissues were incubated at 25°C, 5% CO_2, and 85% relative humidity.

Histological methods. After the desired incubation period, the explants were fixed as follows. Several drops of 4% melted agar were placed on top of the agar block containing the regenerate. After the former solidified, the tissue so enclosed was fixed in Bouin's fixative. Explants as well as control eyes were dehydrated, embedded, sectioned, and stained with hematoxylin and eosin.

Five-, 10-, 15-, and 20-day-old regenerates were studied. Five- and 10-day regenerates were cultured for 5, 10, and 15 days, and 15- and 20-day regenerates were cultured for 5 and 15 days. A total of 94 regenerates were studied; 9 or 10 regenerates per group.

The criteria proposed by Yamada (1967a) were used to determine the developmental stages of the control regenerates. These are based on the system originally proposed by Sato (1940) for *T. pyrrhogaster* and adapted for *T. viridescens* by Stone and Steinitz (1953) and Reyer (1954).

RESULTS

The use of agar as a base for the explants provided a solid support where cell migration from the iris was minimal. In all regenerates cell migration was almost nil even after 15 days in culture. The criteria used to determine whether cell activation continued were the presence of nuclear enlargement with the appearance of nucleoli, the occurrence of depigmentation, cell division, cell elongation, and formation of lens fibers. Throughout this study the explants were compared with the corresponding control regenerate to ascertain whether development continued.

Five-Day Regenerates

Controls for this series were classified as being between stages I and II. In this period activation of the pigmented dorsal iris cells is evidenced by nuclear enlargment and reduction of pigment. The interlaminar space between inner and outer epithelial layers of the dorsal iris becomes conspicuous (Fig. 2). The regenerates cultured for 5 days did not show any significant advancement in regeneration as compared with the controls. In all cases however, depigmentation continued; pigment granules appeared to be extruded from the iris cells directly into the surrounding agar. After 10 days in culture, the signs of nuclear activation became evident in these cells; some cells had two or three prominent nucleoli. The majority of the regenerates showed depigmentation and slight elongation, cells ranging from cuboidal to cylindrical. Furthermore, many

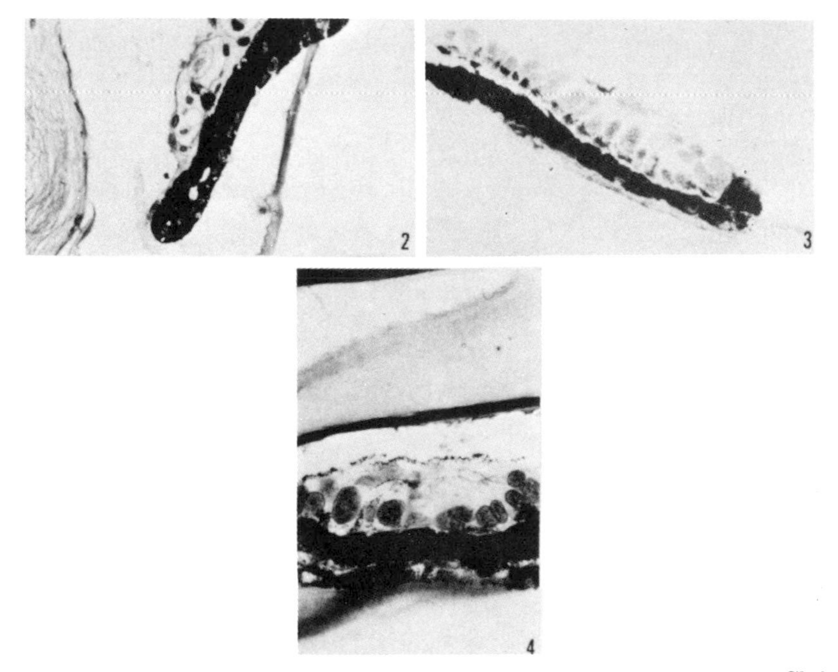

Fig. 2. Representative control stage for the 5 day cultured regenerates. Cleft between inner and outer layers of the dorsal iris is evident. ×180.

Fig. 3. Five-day lens regenerate cultured for 10 days. Agar substrate is at the bottom of the photograph. Cells are partially depigmented. Pigment granules and nuclei have migrated to the base of the cell. ×180.

Fig. 4. Five-day regenerate cultured for 15 days. Cells are completely depigmented, and slight cell elongation has occurred. ×288.

cells had elongated without their being completely depigmented. In these cells nuclei migrated to the base of the cell, together with the pigment granules (Fig. 3). The latter appeared to surround the former. A consequence of the migration of pigment granules to the base of the cells was that the apical end appeared to be free of pigment. These cells had a ruffled free border similar to the cells in active pinocytosis.

Mitotic figures were present in pigmented and depigmented cells though they were few in number. After 15 days in culture some explants developed a depigmented cell population, resembling that of stage IV regenerates (Fig. 4). Mitoses were not present after this period of culture.

Ten-Day Regenerates

Controls for this series ranged between stages early IV to early V. The cleft between internal and external iris layers has enlarged, and a number of depigmented cells are present in the inner wall of the dorsal iris. The latter may become pseudostratified due to an accumulation of cells at all stages of depigmentation (Fig. 5).

A distinctive characteristic of regenerates at these stages was the very high degree of cell proliferation occurring in explants cultured for 5 days. Mitotic figures were abundant and occurred in pigmented and depigmented cells. The general appearance of these explants was that of a mass of depigmented cells undergoing mitosis (Fig. 6). This behavior continued in regenerates cultured for 10 days. In this series one explant showed structures resembling primary lens fibers of the core region of a stage VIII regenerate (Fig. 7). After 15 days in culture, most of the cells present in the explants were depigmented but a few remained full of pigment granules. The cells did not elongate or form fibers in the majority of the explants (Fig. 8). Mitotic figures had disappeared by the fifteenth day in culture.

Fifteen-Day Regenerates

Control regeneration stages for this series were between VI and VII. Depigmented cells of the inner layer of the lens vesicle stop dividing and start to elongate into primary lens fibers. The presumptive lens epithelium and the lens stalk become evident, and these are the sites where the mitotic cell population of the regenerating lens is located (Fig. 9). After 5 days in culture the regenerates showed a number of cells which had elongated into primary lens fibers, developing a structure resembling the lens fiber core of stages IX to X (Fig. 10). The proliferating cell population originally located in the lens stalk and presumptive lens epithelium was usually present at the periphery of the

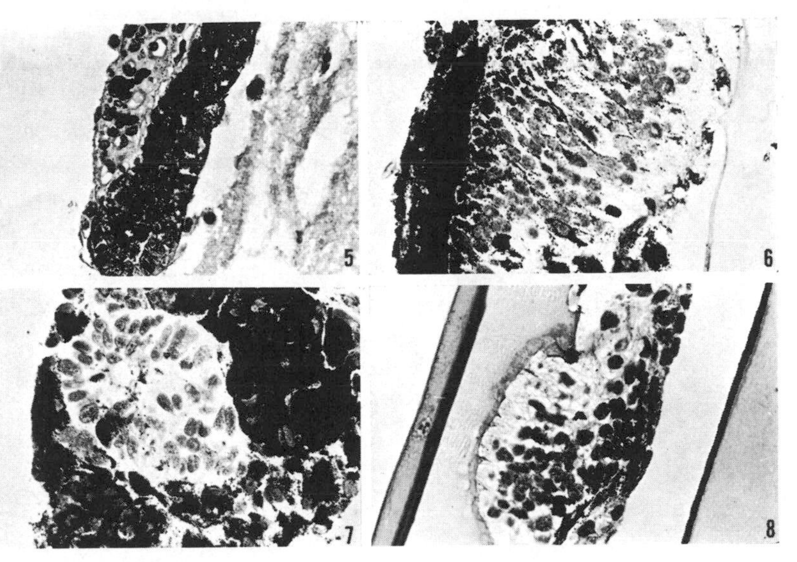

Fig. 5. Representative control for the 10-day cultured regenerates. A number of depigmented cells are present in the inner wall of the dorsal iris. ×180.

Fig. 6. Regenerate, 10 days after lentectomy cultured for 10 days. A large number of depigmented and partially depigmented cells are present. Mitotic figures are abundant in other sections of this regenerate. ×180.

Fig. 7. Another regenerate of the same series as that of Fig. 6. A core region similar to that of a stage VIII regenerate has developed. Mitotic figures are abundant in the peripheral regions of the explant. ×180.

Fig. 8. Ten-day regenerate cultured for 15 days. An abundant depigmented cell population is present, but no lens fiber formation has occurred. ×180.

explant, where it had given rise to an abundant population of depigmented cells. This was evident from the frequent mitotic figures present in this area.

After 15 days in culture many of these regenerates had differentiated fibers corresponding to approximately stage X (Fig. 11). However no lens epithelium was present as such in any of the cultures. A few cases showed a population of depigmented cells at the periphery of the explants. These cells had not elongated.

Twenty-Day Regenerates

Control regenerates for this series ranged from stages IX to X. The primary lens fiber core is almost completely developed, and secondary lens fiber formation is taking place. The presumptive lens epithelium varies from one to two cell layers in thickness, and it is still continuous with the cells of the lens stalk (Fig. 12). The 20-day regenerates cultured

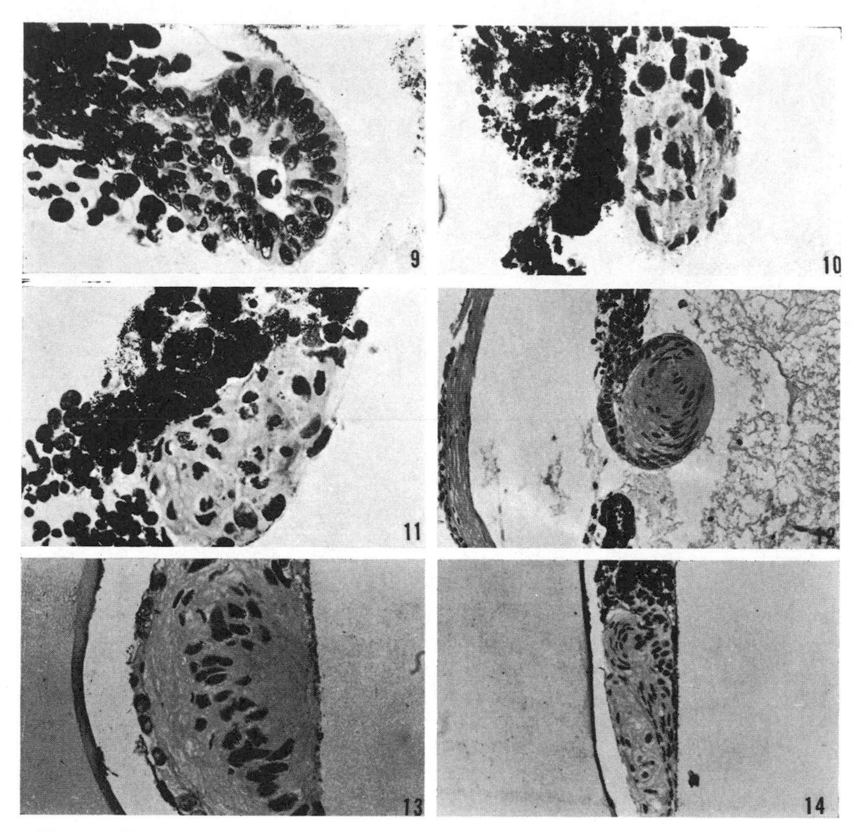

FIG. 9. Representative control stage for the 15-day cultured regenerates. Depigmented cells of the inner layer of the lens vesicle are elongating into primary lens fibers. ×180.

FIG. 10. Fifteen-day regenerate after 5 days in culture. Lens fibers have differentiated in this explant. ×180.

FIG. 11. Regenerate, 15 days after lentectomy cultured for 15 days. A lens fiber core is present, but no lens epithelium has developed. ×180.

FIG. 12. Control for the 20-day cultured regenerates. The primary lens fiber core has developed and secondary lens fibers are differentiating. ×72.

FIG. 13. Twenty-day regenerate cultured for 5 days. Secondary lens fibers have continued differentiating. The lens epithelium is flattening to become squamous in type. ×180.

FIG. 14. Twenty-day regenerate cultured for 15 days. Lens fibers have oriented around secondary centers of elongation. No lens epithelium is present. ×72.

for a 5-day period continued to differentiate and attained stages XI to XII. Mitotic figures were present in pigmented as well as depigmented cells, but these cells did not elongate. The lens epithelium either flattened into a squamous type cell, or elongated into fibers (Fig. 13). Secondary lens fibers continued to differentiate; in many cases they oriented around a secondary center of elongation which was established perpendicularly to the original primary lens fiber core (Fig. 14). After 15 days in culture these regenerates developed into lentoids in which the development of primary and secondary lens fibers correspond to that of a mature lens. These regenerates, however, did not undergo a harmonious differentiation, but were flattened and had an elipsoidal shape. Lens epithelium was not present as such in any of the explants.

From the foregoing experiments, development of the lens regenerates under the *in vitro* conditions employed may be summarized as follows. Once nuclear activation appears and reduction of pigment granules has started in the dorsal iris (stages I, II) the cells continue to undergo nuclear activation and depigmentation. They give rise to a number of depigmented cells capable of elongating into a cuboidal or cylindrical shape. The regenerates can proceed to stage III with extensive depigmentation, but they are not capable of forming lens fibers.

After depigmentation is accomplished in some cells (stages III to IV), the cells are capable of entering mitosis and proliferating. These cells give rise to a depigmented cell population that ceases to divide after 15 days in culture. They remain depigmented and are not capable of differentiating into lens fibers in the majority of cases. After depigmentation is completed and cell elongation begins (stages VI to VII), some fibers differentiate under culture conditions. When primary lens fibers have developed and secondary lens fibers have started to differentiate (stages IX to X), regenerates develop *in vitro* into structures resembling adult lenses. In these cases lens epithelial cells do not remain as a dividing proliferating population, but elongate and form lens fibers. Secondary lens fibers do not develop around the original primary lens fiber core, but orient around secondary nuclei of elongation.

DISCUSSION

There is limited information regarding *in vitro* development of lens regenerates although a number of investigators are engaged in the study of this phenomenon. The first attempt to culture the dorsal iris *in vitro* was that of Stone and Gallagher (1958). The normal iris membrane of nonlentectomized newts was cultured on pieces of rayon acetate in a medium composed of horse serum and chick embryo extract. The authors reported absence of lens regeneration after 24 days in culture. A few iris membranes, however, were capable of giving rise to lens tissue when transplanted into a lentectomized eye. More recently Eguchi (1967), working with *T. pyrrhogaster*, cultured the iris membrane and the iridocorneal complex on solid media composed of agar, Tyrode solution, chick embryo extract, and horse serum. He succeeded in culturing the regenerating lens only as part of the iridocorneal complex. Cultures of the isolated iris ring showed suppression of lens regeneration, followed by disintegration of the iris tissue. Eisenberg-Zalik and Meza (1968) reported that maintenance of a CO_2 atmosphere was essential for the optimal development of regenerates. It may be that the absence of regeneration in the isolated dorsal iris (Eguchi, 1967) was due to lack of CO_2 atmosphere. In general our findings, based on the culture of isolated dorsal iris, agree in several aspects with those of Eguchi (1967), based on culturing the lens as part of the iridocorneal complex, and with the preliminary results of Reese as reported by Yamada (1967a). After nuclear activation has occurred (stages I to II), depigmentation continues. Following depigmentation (stages III to IV), the cells are capable of proliferating *in vitro* but remain undifferentiated. After cells start elongating (stages V and onward), further progress of differentiation can be obtained *in vitro*.

From the results obtained in this study one can distinguish three stages in the transformation of iris cells into lens cells: depigmentation, multiplication, and elongation. Whether specific factors in the ocular environment responsible for inducing each of these stages can be reproduced *in vitro* remains to be established. In this context some preliminary results obtained by these authors seem to be of interest. Ten-day regenerates which give rise to a depigmented cell population at 5% CO_2 atmosphere, differentiate primary lens fibers at higher CO_2 concentrations. A similar situation has been reported by Morton (1967) for erythrocyte maturation where higher CO_2 concentrations, up to 30%, stimulated reticulocyte and erythrocyte formation from immature marrow cells. This situation suggests that factors simulating conditions for cell differentiation may be developed in *in vitro* systems.

The observation that, once depigmented cells begin to elongate, they can continue fiber formation *in vitro* is significant. A similar finding based on *in vivo* studies was reported by Ikeda (1936) and Ikeda and Amatatu (1941). They found that the regenerating lens continued its development in the fourth brain ventricle of *Hynobius* larva, only if it was transplanted after the lens vesicle stage.

Appearance of tissue-specific proteins is first detected by the fluorescent antibody technique in the elongated cells of the inner cell layer of stage IV regenerates (Takata *et al.*, 1964). It appears that once cells stop synthesizing DNA (Eisenberg and Yamada, 1966) and start producing lens-specific proteins, they are capable of differentiating a lens fiber *in vitro*. A long-lived messenger RNA is synthesized in the developing lens of the chick embryo, (Reeder and Bell; Scott and Bell 1965); and the bovine lens epithelium (Papaconstantinou, 1967; Stewart and Papaconstantinou 1967). Whether a similar situation exists in the regenerating lens remains to be established.

SUMMARY

The development *in vitro* of the regenerating lens at successive time intervals after lens removal was studied in the adult newt *Triturus viridescens*. Regenerates were cultured using an agar substrate in which a synthetic medium with serum supplement was incorporated. Explants were incubated at 25°C, 5% CO_2, and 85% relatively humidity, and cultured for 5, 10, and 15 days.

Regeneration stages I to II, in which nuclear activation and depigmentation are starting to occur, continue to extrude their pigment granules and give rise to a population of completely depigmented cells. After depigmentation has occurred *in vivo* in some cells of the dorsal iris (stages III to IV), a depigmented cell population is produced capable of limited proliferation but incapable of lens fiber differentiation. When cell elongation has occurred in the depigmented cells of the inner wall of the regenerating lens (stages V and onward), explants are capable of developing lens fibers *in vitro*. These results are discussed and related to the available information on RNA and protein synthesis in the lens regenerating system.

REFERENCES

Eguchi, G. (1963). Electron microscopic studies on lens regeneration. I. Mechanisms of depigmentation of the iris. *Embryologia* **8**, 45–62.

Eguchi, G. (1964). Electron microscopic studies on lens regeneration. II. Formation and growth of the lens vesicle and differentiation of lens fibers. *Embryologia* **8**, 247–287.

Eguchi, G. (1967). *In vitro* analysis of Wolffian lens regeneration. Differentiation of the regenerating lens rudiment of the newt, *Triturus pyrrhogaster*. *Embryologia* **9**, 246–266.

Eisenberg, S., and Yamada, T. (1966). A study of DNA synthesis during the transformation of iris into lens in the lentectomized newt. *J. Exptl. Zool.* **162**, 353–368.

Eisenberg-Zalik, S., and Meza, I. (1968). *In vitro* culture of the regenerating lens. *Nature* **217**, 179–180.

Eisenberg-Zalik, S., and Yamada, T. (1967). The cell cycle during lens regeneration. *J. Exptl. Zool.* **165**, 385–394.

Ikeda, Y. (1936). Neue Versuche zur Analyse der Wolffschen Linsenregeneration. *Arb. Anat. Inst. Kaiserl. Japan. Univ. Sendai* **18**, 1–16.

Ikeda, Y., and Amatatu, H. (1941). Über den Unterschied der Erhaltungsmöglichkeit der Linse bei zwei Urodelenarten (*Triturus pyrrhogaster* und *Hynobius nebulosus*), die sich bezüglich der Fähigkeit zur Wolffschen Linsenregeneration voneinder wesentlich verschieden verhalten. *Japan. J. Med. Sci.* (*I. Anat*) **8**, 205–226.

Karasaki, S. (1964). An electron microscopy study of Wolffian lens regeneration in the adult newt. *J. Ultrastruct. Res.* **11**, 246–273.

Morton, H. (1967). Role of carbon dioxide in erythropoiesis. *Nature* **215**, 1166–1167.

Papaconstantinou, J. (1967). Molecular aspects of lens cell differentiation. *Science* **156**, 338–346.

Reeder, R., and Bell, E. (1965). Short and long-lived messenger RNA in embryonic chick lens. *Science* **150**, 71–72.

Reyer, R. W. (1954). Regeneration of the lens in the amphibian eye. *Quart. Rev. Biol.* **29**, 1–46.

Reyer, R. W. (1962). Regeneration in the amphibian eye. *In* "Regeneration" (D. Rudnick, ed.), pp. 211–265. Ronald Press, New York.

Sato, T. (1940). Vergleichende Studien über die Geschwindigkeit der Wolffschen Linsenregeneration bei *Triton taeniatus* und bei *Diemyctilus pyrrhogaster*. *Arch. Entwichlungsmech. Organ.* **140**, 570–613.

Scott, R. B., and Bell, E. (1965). Messenger RNA utilization during development of chick embryo lens. *Science* **147**, 405–407.

Stewart, J. A., and Papaconstantinou, J. (1967). Stabilization of RNA templates in bovine lens epithelial cells. *J. Mol. Biol.* **29**, 357–370.

Stone, L. S. (1960). Regeneration of the lens, iris and neural retina in a vertebrate eye. *Yale. J. Biol. Med.* **32**, 464–473.

Stone, L. S., and Gallagher, S. B. (1958). Lens regeneration restored to iris membranes when grafted to neural retina environment after cultivation *in vitro*. *J. Exptl. Zool.* **139**, 247–261.

Stone, L. S., and Steinitz, H. (1953). Regeneration of lenses in eyes with intact and regenerating retina in adult *Triturus viridescens*. *J. Exptl. Zool.* **124**, 435–468.

Takata, C., Albright, J. F., and Yamada, T. (1964). Lens antigens in a lens regeneating system studied by the immunofluorescent technique. *Develop. Biol.* **9**, 385–397.

Takata, C., Albright, J. F., and Yamada, T. (1966). Gamma crystallins in Wolffian lens regeneration demonstrated by immunofluorescence. *Develop. Biol.* **14**, 382–400.

Yamada, T. (1966). Control of tissue specificity. The pattern of cellular synthetic activities in tissue transformation. *Am. Zoologist* **6**, 21–31.

Yamada, T. (1967a). Cellular and subcellular events in Wolffian lens regeneration. *In* "Current Topics in Developmental Biology" (A. Monroy and A.A. Moscona, eds.), Vol. 2, pp. 247–283. Academic Press, New York.

Yamada, T. (1967b). Cellular synthetic activities in induction of tissue transformation. *Ciba Found. Symp. Cell Differentiation* pp. 116–126. Little, Brown, Boston, Massachusetts.

YAMADA, T., and KARASAKI, S. (1963). Nuclear RNA synthesis in newt iris cells engaged in regenerative transformation into lens cells. *Develop. Biol.* **7**, 595–604.

YAMADA, T., and TAKATA, C. (1963). An autoradiographic study of protein synthesis in regenerative tissue transformation of iris into lens in the newt. *Develop. Biol.* **8**, 358–369.

The *in Vitro* Development of Lens from Cornea of Larval *Xenopus laevis*

J. C. Campbell and K. W. Jones

INTRODUCTION

Of the various *in vivo* metaplastic transformations shown to occur in the eye tissues of several amphibian species (e.g., see Sato, 1951; Stone, 1955, 1958, 1960; Reyer, 1956), perhaps the best known is the ability to regenerate a lens after the removal of the original lens from the eye. In most of such species, including those of the genus *Triturus*, the regenerating lens develops from cells in the pupillary margin of the dorsal iris (see review, Reyer, 1954). However, at least two other tissues have been shown to act as a source for the regenerate. Neural retina can give rise to lens tissue as has been shown in *Salamandra salamandra* by Fischel (1903), in *Ambystoma mexicanum* by Törö (1932), and in *Xenopus laevis* by Campbell (1963). The role of the cornea in providing a regenerated lens has been demonstrated in *Hynobius unnangso* (Ikeda, 1936, 1939) and in *X. laevis* (Overton and Freeman, 1960; Freeman and Overton, 1961, 1962; Freeman, 1963; Campbell, 1963).

Fewer results have been reported of transformation of eye tissues *in vitro*, though tissues of embryonic chick eyes have been shown to undergo metaplasia both when grafted to the body wall, e.g., tapetum to retina (Alexander, 1937) and in culture, e.g., tapetum to retina and lens, retina to tapetum (Dorris, 1938), and retina to lens (Moscona, 1957). Of the amphibian eye tissues known to regenerate a lens, the iris of *Triturus v. viridescens* has been isolated in tissue culture by Stone and Gallagher (1958), but no transformation into lens tissue occurred. If the iris was subsequently transferred to a lentectomized eye, a lens was regenerated from its dorsal pupillary margin. It was concluded from these results that the presence of neural retina is a prerequisite for lens regeneration in this species.

In *Xenopus laevis,* Balinsky (1951) has shown that during normal embryogenesis a lens can develop from the head ectoderm in the absence of an eyecup, i.e., free lens formation. Thus, although Freeman (1963) suggests that the eyecup must be present during regeneration of the lens from the cornea, it seemed that the cornea might have the capacity to transform to lens tissue when isolated from the eye in culture conditions. Accordingly, the experiments to be described were undertaken to determine the metaplastic activity of larval *X. laevis* corneal tissue *in vitro.*

MATERIAL AND METHODS

Animals. Larval *X. laevis* from induced matings were used between developmental stages 55 and 58 (Nieuwkoop and Faber, 1956). The animals had been reared in tap water and fed on dried nettle powder.

Operation. The animals were anesthetized in a 1/2000 solution of M.S. 222 (Sandoz) and then passed through three baths of sterile water to remove gross contamination. The operation was carried out in sterile water. An incision was made at the anterior margin of the cornea and a flap was cut posteriad to include the region of the cornea overlying the pupil of the eye. The piece of tissue, which often contained some pericorneal ectoderm from the region immediately posterior to the eye, was removed and transferred to Barth's saline (Barth and Barth, 1959) prior to being set up in culture. Due to the method of excision and since, at the stage of larvae used, the outer and inner corneas are incompletely fused, the tissue removed consisted almost wholly of outer cornea.

Culture methods. The culture chambers used, described previously by Jones and Elsdale (1963), consist of a glass slide, 3 × 1 × ⅛ inches with a ⅝-inch hole bored through the center, the bottom of which is sealed by a ⅞-inch coverslip waxed into place. After the tissue and medium are introduced to the well, the top is similarly sealed, excluding air to facilitate microscopic examination. The chambers have an approximate capacity of 0.3 ml.

The excised corneal tissue was transferred from the Barth's saline to small drops of fowl plasma previously placed in the bottom of each chamber. Normally two pieces of tissue were placed in each chamber. After clotting of the plasma, the chambers were filled with Morgan,

Morton, and Parker's Medium 199 diluted to half strength with sterile deionized water and subsequently sealed. The tissues obtained sufficient oxygen for normal metabolism from that present in the medium; the pH, as indicated by phenol red, was controlled by changing the medium at approximately 3-day intervals. Slight modifications in the technique are described in the relevant sections.

The cultures were examined daily with a Cooke, Troughton, and Simm's inverted phase-contrast microscope and photographed on Ilford Pan F film.

Immunological methods. Antisera to adult *X. laevis* lens were prepared in rabbits aged between 9 and 18 months. Each rabbit received at least ten subcutaneous injections, each containing 15–20 mg of protein, over a period of 6 weeks, and was bled 10 days after the last injection.

The presence or absence of lens antigens in the cultured explants was determined by two methods. In the first, the precipitin ring test (see Kabat and Mayer, 1961), soluble extracts were prepared by homogenizing together at least six explants of the appropriate stage. The homogenate was centrifuged and the supernatant was adjusted to a protein concentration of approximately 15 mg/ml as estimated by a Bellingham and Stanley pocket refractometer. The adjusted solution was layered over an aliquot of antilens serum in a semicapillary tube, allowed to incubate for 1 hour at room temperature, and then checked for the presence or absence of a precipitate at the interface. Negative reactions were allowed to incubate for a further hour and then rechecked. The presence of a precipitate at the interface after either 1 or 2 hours indicated the presence of lens antigens in the culture extract.

The second method used was the immunofluorescent technique. Anti-*X. laevis* lens antiserum was fractionated by 40% saturation of ammonium sulfate. The globulins thus precipitated were coupled to rhodamine sulfonyl chloride prepared from lissamine rhodamine B 200 (I.C.I. Ltd.). Uncoupled fluorochrome was removed from the globulin solution by passage through Sephadex G-25 (Pharmacia) and subsequent dialysis against phosphate-buffered saline, pH 7.2, of the first eluted fraction.

The cultures to be examined were snap frozen in trichlorofluoromethane (Arcton 12, I.C.I. Ltd.) cooled by liquid oxygen to −183°C. The subsequent fixation was by freeze substitution in alcohol at

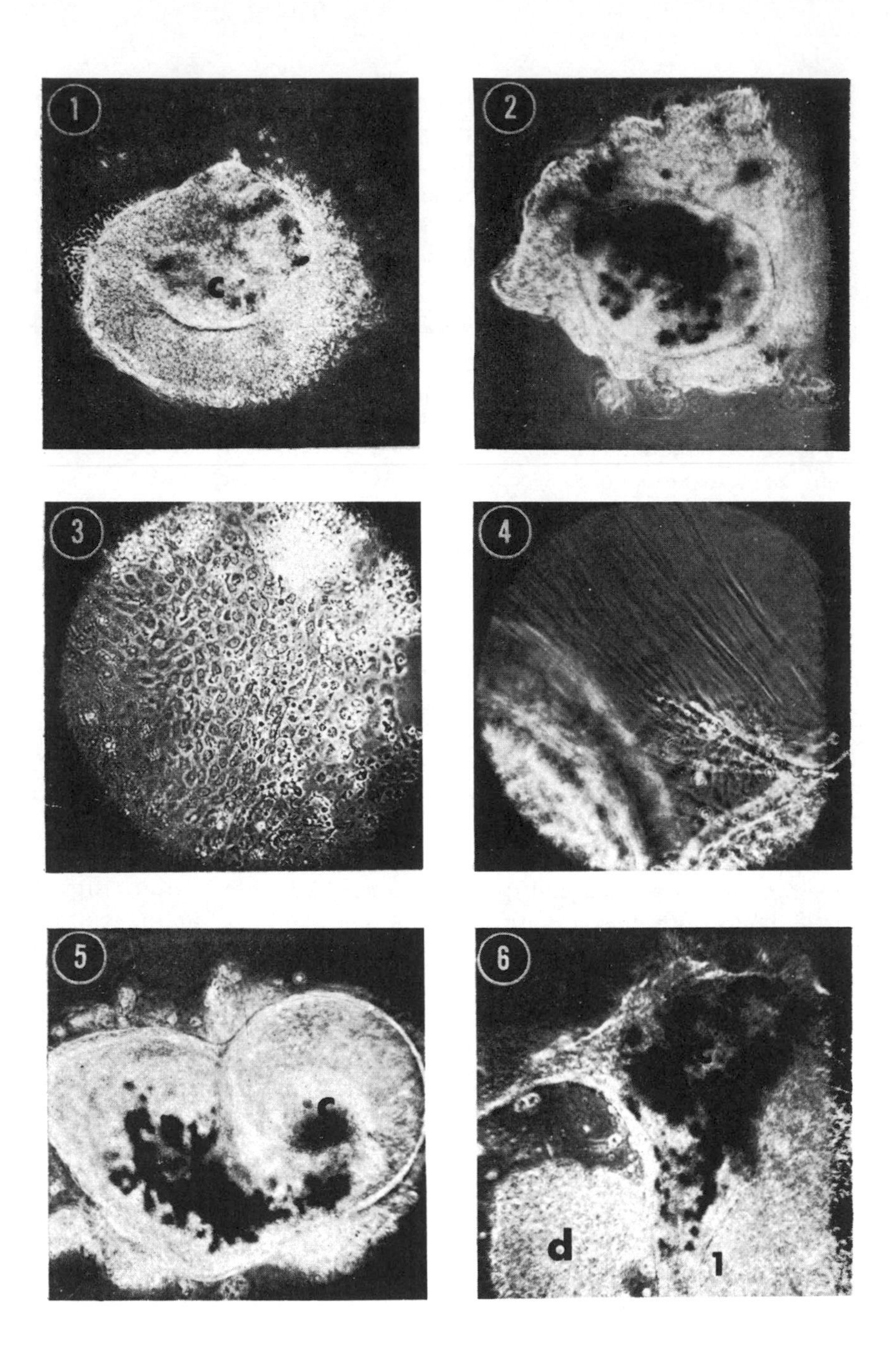
1
c
2
3
4
5
c
6
d
1

—76°C. The cultures were embedded in paraffin wax and 5-μ sections were cut. The sections were allowed to react with labeled antibody for 30 minutes at room temperature, washed for at least 30 minutes in several changes of phosphate-buffered saline, pH 7.2, and then examined for fluorescence under a Reichert Zetopan microscope incorporating a HBO-200 mercury vapor lamp and using primary filters (Schott) UG 1 and BG 12 and secondary filter GG 9. The controls performed were similar to those previously described (Campbell, 1965).

OBSERVATIONS AND RESULTS

Explants Containing Cornea and Pericorneal Ectoderm (50 Explants)

This group, which was the largest studied, consisted of explants which had been dissected without any precaution to exclude pericorneal ectoderm. However, the amount of ectoderm in any explant never exceeded 20% of the mass of cornea. The initial appearance of the explant was a homogeneous sheet of polygonal cells without any obvious morphological variation. At first the sheets presented a crumpled appearance and the edges were irregular, but within hours of explantation the edges rounded off, leaving behind damaged cells.

Within the first day after explantation the most striking change was the appearance of a well-defined condensation of elongated cells forming a dense fibrous mass and often surrounded by a capsule of cells

---

Key to abbreviations: *c*, condensate; *d*, disintegrating cells.

FIG. 1. Living culture, phase contrast. Explant of cornea and pericorneal ectoderm after 1 day in culture showing appearance of condensate. × 45.

FIG. 2. Living culture, phase contrast. Same as Fig. 1, demonstrating association of pigment cells with developing condensate. × 50.

FIG. 3. Living culture, phase contrast. Outgrowth of polygonal cells from edge of 1-day explant of cornea and pericorneal ectoderm. × 420.

FIG. 4. Living culture, phase contrast. Fibrils appearing from edge of 1-day explant of cornea and pericorneal ectoderm. × 420.

FIG. 5. Living culture, phase contrast. Explant of cornea and pericorneal ectoderm after 2 days in culture. The condensate has assumed a spherical appearance. × 45.

FIG. 6. Living culture, phase contrast. Part of explant of cornea and pericorneal ectoderm after 3 days in culture. The association of pigment cells with the condensate is marked, and the initial stages in the disintegration of the cells surrounding the condensate can be seen. × 90.

which appeared bright in phase contrast (Figs. 1 and 2). In a few cases more than one condensation appeared within the explant, but in such cases the condensations normally fused on the second or third day of culture. Closely associated with the condensates were pigment cells which had been included in the explant. These normally migrated into the region surrounding the condensate and then rounded up, although in a few cases they migrated into the condensate itself.

During this initial period the edges of the explant showed outgrowth, which, though limited in extent, appeared in a great variety of forms, indicating that there may be a great potential for developmental divergence among corneal cells. Several distinctive types of outgrowth were found regularly. One took the form of coherent sheets of polygonal cells closely resembling the original cell type (Fig. 3). These cells were characterized by a perinuclear ring of phase-dense cytoplasm, and relatively small dense nucleoli. Since mitoses were never observed within such sheets, they presumably arose from migration rather than cellular proliferation. A second type of outgrowth had the appearance of thick cytoplasmic processes within which no nuclei could be discerned. Such structures rarely persisted, but withdrew leaving fragments of cytoplasm behind. A further type appeared as tufts of fine, rigid-looking fibrils which fanned out into the plasma from sites on the explant border closest to the condensate (Fig. 4). That such structures were not merely stress lines in the plasma caused by shrinkage or by change in the shape of the explant is suggested by the subsequent emergence from the explant of cells with identical processes. Spiky, phase-dense, cytoplasmic processes, closely resembling nerve cell growth cones were also seen.

During the second day of culture the condensates became larger and assumed a spherical or subspherical appearance (Fig. 5), but little change was seen in the rest of the explant. However, from the third day after explantation onward a considerable change was seen in behavior of the majority of cells not associated with the condensate. Extensive outgrowth and migration was seen leading to the formation of a hemispherical dome of cells, the condensate being included as part of the wall. Viewed in transillumination the structure had the appearance of an expanding ring, since the mass of cells in the center of the explant gradually rounded up and disintegrated (Figs. 6–9). In some cases, however, the intact cellular connections were maintained between the condensate and parts of the expanding wall of the ring (see, e.g., Figs. 7 and 8). Several dead cells were normally found within the hemisphere, and, if the chamber was inverted, these could be seen to fall freely through the enclosed fluid, thus demonstrating that the space enclosed by the dome contained no clotted plasma. The main structure of the dome was formed by a network of cells which contained a high density of filamentous structures. Other cell types, for example, spindle-shaped fibroblast-like cells, epithelial sheets and clumps of rounded cells, were associated with the main network.

In some cases when the condensate formed centrally within the mass of the explant, it came to lie within the dome, maintaining several cellular strandlike connections with the walls. As the dome expanded the connections became more attenuated and the weaker ones were lost.

In the cases observed there was no apparent limit to the expansion of the dome, which continued until the cells lost contact with one another, thus destroying the integrity of the explant. Subsequently the individual cells and small groups of cells migrated randomly. In many cases the condensate remained as a smooth mass of highly refractile cells, which, when manipulated, had an unyielding consistency.

Explants of Cornea Alone (18 Explants)

This group of experiments was carried out in an identical manner to that described for the previous group, except that the explants contained cornea alone. Apart from a lower incidence of morphological change, the events observed were identical to those described in the previous section.

Explants of Pericorneal Ectoderm Alone (18 Explants)

Pericorneal ectoderm, similarly cultured, did not reproduce the events already described, but behaved like cultured ectoderm from other regions of the body, giving rise to sheets of epithelial and fibroblastic cells.

Combined Cultures of Cornea and Pericorneal Ectoderm (24 Explants)

In these experiments the cornea was separated from its surrounding ectoderm and the two tissues were cultured in close proximity in the same plasma clot. In those instances where the two tissues established contact, both contributed to the formation of the dome, whereas

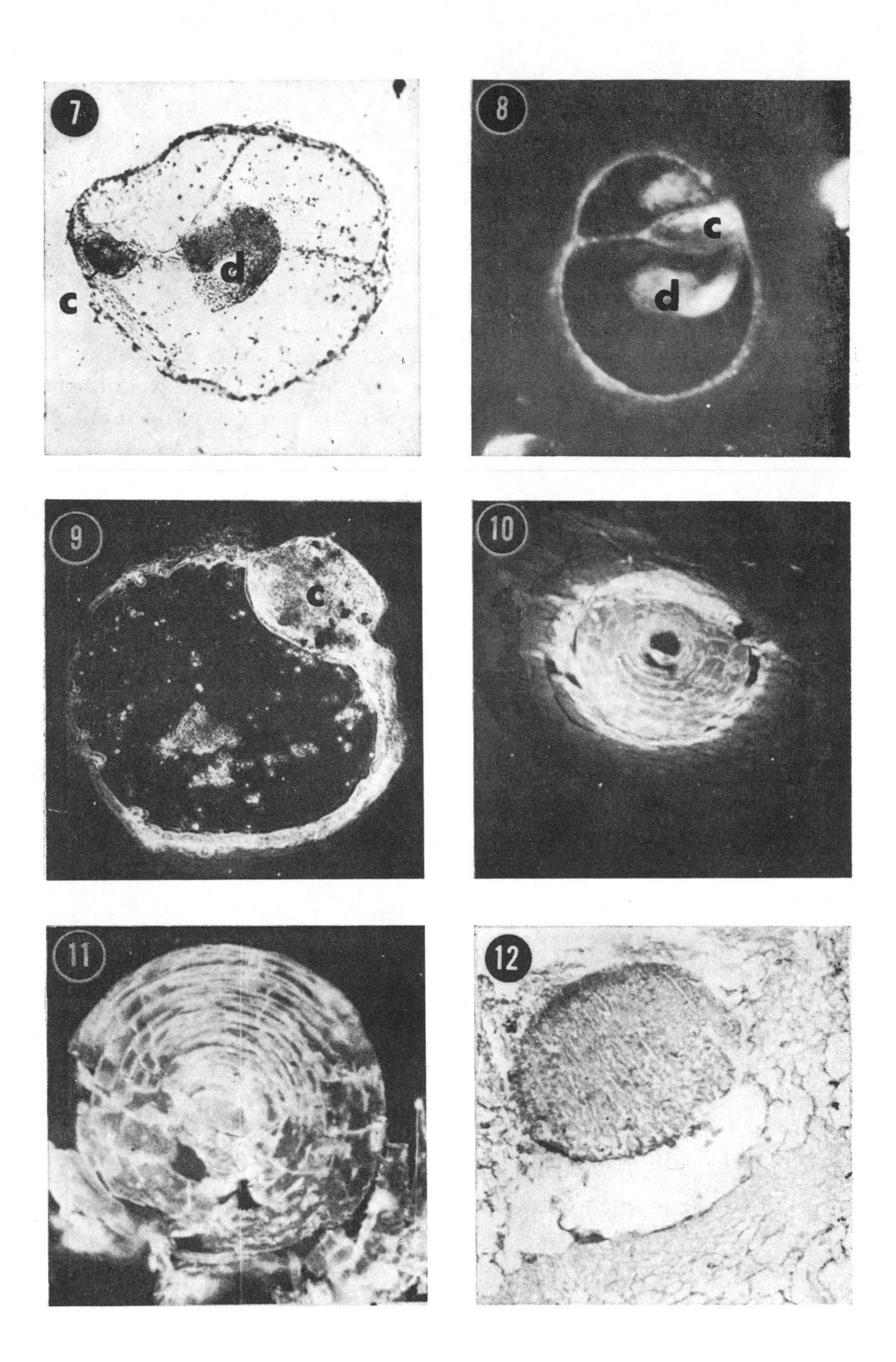
7
c
d
8
c
d
9
c
10
11
12

if contact was not established the cornea alone formed a dome. The incidence of dome formation was highest when the explants fused, almost every case showing the typical morphological changes.

Variations in Culture Conditions

In order to assess the effect of the culture conditions on the expression of the condensate and dome formation, the following experiments were undertaken, using explants containing both cornea and pericorneal ectoderm. When a 2% solution of gelatin in the culture medium was substituted for plasma as the supporting medium (30 explants) there was a marked fall in the incidence of the dome phenomenon, though in a few cultures it was fully expressed. Omission of a supporting substrate (30 explants) led to the formation of hollow vesicles with local thickenings in the wall, which may correspond to the condensates of the other cultures.

Immunological Results

Precipitin ring tests were carried out on explants containing both cornea and pericorneal ectoderm at daily intervals from the first day after explantation to the seventh day. Samples were chosen from the

---

FIG. 7. Living culture, phase contrast. Explant of cornea and pericorneal ectoderm after 5 days in culture. The peripherally located condensate retains connections with other areas of the wall of the hemisphere by means of cellular strands. The mass in the center of the explant consists of degenerating cells. × 24.

FIG. 8. Living culture. Similar explant to that of Fig. 7. Viewed by epi-illumination. Two masses of disintegrating cells lie adjacent to the condensate. × 30.

FIG. 9. Living culture, phase contrast. Explant of cornea and pericorneal ectoderm after 7 days in culture. Loss of cells from center of explant is virtually complete. × 18.

FIG. 10. Fixed and sectioned, immunofluorescent technique. Lentoid from 7-day culture of cornea and pericorneal ectoderm treated with rhodamine-labeled anti-*X. laevis* lens serum The labeling is restricted to the central area of the structure. × 180.

FIG. 11. Fixed and sectioned, immunofluorescent technique. Similar explant to that of Fig. 10, treated with rhodamine-labeled anti-*X. laevis* lens serum. Labeling is much more irregular than in the previous figure. × 240.

FIG. 12. Fixed and sectioned, stained with Delafield's hematoxylin and eosin. Similar explant to those of Figs. 10 and 11. The histological appearance of the lentoid is atypical, with the fibers oriented in the anteroposterior plane rather than concentrically. × 180.

series which had been cultured in a plasma clot and from the series without supporting substrate. With both types of culture, lens antigens were not detectable until the fourth day after explantation but were found on each subsequent day.

Immunofluorescent tests were carried out on explants which had been cultured for 7 days in plasma-clot. The localization of fluorescence showed that lens antigens were present only in the condensate, though not uniformly throughout it (Figs. 10 and 11). In general, the central region of the condensate was strongly and uniformly labeled, but the more peripheral areas either lacked fluorescence or showed the reaction only in scattered patches. The morphological appearance of the condensates was in many cases very similar to that of a normal lens with concentrically arranged layers of fibers and an epithelium over part of the surface. In some cases (e.g., Fig. 12), the condensate, although composed of fibrous elements, did not have them arranged concentrically. Conventional histology showed the condensates to be strongly eosinophilic, a feature characteristic of lens tissue.

DISCUSSION

The observations and experiments which have been described indicate conclusively that the cornea of larval *X. laevis* can transform into a lens when isolated from the rest of the eye *in vitro*. The region of the cornea which will transform into lens becomes distinguishable from the rest of the culture at an early stage and subsequently grows and assumes a lentoid shape. Although the gross appearance of the lens was in most cases normal, a few instances were found in which the appearance of the lentoid was abnormal. Such cases, which can also be found during lens regeneration from the cornea *in vivo* (Campbell, unpublished), had, however, the normal histological components of a lens, i.e., fibers and an epithelium.

In some explants the preliminary condensations were found in more than one area, though these eventually coalesced to form only one lentoid. Freeman (1963) has shown that *in vivo* the cytological changes which are found in the cornea prior to any morphological sign of lens formation, i.e., change in nucleolar number, may be found in more than one region of the cornea, but subsequently only one such region forms the regenerate. Thus it is clear that the potential for lens formation is distributed throughout the cornea and is not restricted to one area alone. These observations do not indicate, however, whether the whole corneal cell population has this potential, or whether it is restricted to several discrete subpopulations.

Apart from the formation of the lentoid, the relative movements and behavior of the explant as a whole show that a well-defined morphogenesis is occurring leading to the formation of the hemispherical dome. With the exception of the dead cells mentioned, the inner contents of the dome are liquid and it may be that the pressures exerted by this fluid lead to the formation of the typical structure, in this way resembling the morphogenetic forces involved in the formation of the spatial relationships of the normal eye, including corneal size and curvature (Coulombre, 1956, 1957; Coulombre and Coulombre, 1958).

Although it has been shown that *X. laevis* cornea contains at least one antigen in common with the lens, the antisera used in the experiments described showed no cross-reaction with normal cornea. Thus the appearance of lens antigens in the explants after 4 days in culture can be considered as a genuine change in the biochemical nature of the tissue. Using the same antisera, it has been shown by immunofluorescence (Campbell, 1965) that during regeneration of the lens from cornea *in vivo* lens antigens become detectable in the regenerating tissue 24 hours after lentectomy. This difference in time of appearance of the antigens may be real or may be dependent on the fact that different techniques were used for the two sets of observations. The immunofluorescent studies described above have shown that the distribution of lens antigens in the lentoid is not uniform, but is restricted mainly to the central region, a situation similar to that found in some lenses regenerated *in vivo* (Campbell, 1965).

It is clear that isolated cornea has the ability to effect the transformations described, though the expression of the change is apparently reinforced by the presence of pericorneal ectoderm in contact with the cornea in the culture. The pericorneal ectoderm alone, however, was unable to form lens, behaving similarly to ectoderm from other regions of the body in forming sheets of epithelium and fibroblasts. Freeman (1963) has shown that *in vivo* the pericorneal ectoderm can form a lens when it has been allowed to replace the epithelium of the outer cornea. This difference can best be explained by supposing that the ectoderm, *in vivo,* had first been transformed into cornea, presumably under the influence of the eyecup (see Lewis, 1905) and that this in turn can form lens. *In vitro* it may be that the pericorneal ectoderm cannot form cornea and thus cannot transform into lens.

Freeman (1963) has shown that the presence of the eyecup is necessary for the regeneration of the lens from the cornea *in vivo*. However, such a condition is obviously not necessary *in vitro*. This discrepancy raised the question of whether some factor in the culture environment was replacing the influence of the eyecup on the cornea. However, since it was shown that a lens can be formed in culture with either plasma or gelatin as the supporting medium or even with no supporting medium, it seems likely that cornea, of itself, possesses the capacity to transform into lens in the absence of other lens tissue. It should be pointed out though, that alterations in the nutritive fluid medium were not made. One possible explanation of the difference between Freeman's findings and those reported here could be the difference in the availability of nutrients to the cornea in the two cases. In Freeman's case removal of the eyecup involves removal of the aqueous humor, which is partially responsible for providing nutrients to the cornea. Thus the metabolism of the corneal tissue must be disturbed and this may inhibit transformation, whereas in culture nutrients are readily available, thus perhaps facilitating transformation. Balinsky's (1951) finding of free lens formation in *X. laevis* embryos is consistent with such a hypothesis since the head ectoderm of the early *Xenopus* embryo has a plentiful supply of yolk (Selman and Pawsey, 1965). Thus any capacity of the ectoderm to transform into lens will not be inhibited by lack of an adequate supply of nutrients.

The precise nature of the initial stimulus to transformation is still obscure, though from the results described, it must involve a derepression acting directly on the cornea. However, whether this is the removal of a positive repression, i.e., some substance liberated, presumably from the lens, which inhibits corneal change, or of a negative repression, i.e., preferential removal of a factor necessary for regeneration, by some other component of the eye system, again presumably the lens, is not clear. If the repression is negative it must mean that the appropriate factor or factors are present in the culture environment. If the necessary factors are simply amino acids and other growth requirements, as postulated by Takano (1958), then such a condition may be fulfilled. If, however, they are more specific inducers it seems unlikely that they would be present in the culture conditions and the repression is more likely to be positive.

Although the demonstration of the formation of lens in culture has been the main concern of these experiments, the behavior of the other cells of the explant exhibits some interesting features. The pigment cells included in the explant initially show a high migratory activity, tending to move to the area of subsequent lens formation, but once the condensation appears the migration is inhibited. The inhibited pigment cells round up, but do not apparently lose their pigment granules nor do they have any obvious role in the formation of the lens. Such changes are not merely a reflection of the effect of the culture conditions, since, in the few cases where lens formation did not occur, the pigment cells remained normally migratory and obviously viable.

During the period of culture several cell types become obvious in the explant, indicating that the cells of the cornea may have considerable developmental potential, which is not normally expressed. The appearance of these cells may be a consequence of the general derepression of the cornea, but what additional factors and conditions are necessary for the establishment of new developmental pathways was not investigated. However, the system described seems a suitable one with which to investigate factors involved in developmental stability.

SUMMARY

Corneas from larval *Xenopus laevis,* stages 55–58, were set up in plasma-clot tissue culture with Medium 199. Initially, a condensate of fibrous cells appeared within the explant. Over the course of 7 days the condensate transformed into a lentoid, which in many cases had a histologically normal appearance. The remainder of the explant developed into a fluid-filled hemisphere enclosing the lentoid. This typical morphogenetic pattern occurred more frequently when pericorneal ectoderm was included in the explant in contact with the cornea, but was not reproduced by pericorneal ectoderm alone nor by ectoderm from any other part of the body.

Lens antigens appeared in the explants after 4 days in culture, and the final distribution of lens antigens in the lentoid is described.

The appearance and behavior of other cell types in the explant are also reported.

REFERENCES

ALEXANDER, L. E. (1937). An experimental study of the role of optic cup and overlying ectoderm in lens formation in the chick embryo. *J. Exptl. Zool.* **75,** 41–74.

BALINSKY, B. I. (1951). On the eye cup-lens correlation in some South African amphibians. *Experientia* **7,** 180–181.

BARTH, L. G., and BARTH, L. J. (1959). Differentiation of cells of *Rana pipiens* gastrula in unconditioned medium. *J. Embryol. Exptl. Morphol.* **7,** 210–222.

CAMPBELL, J. C. (1963). Lens regeneration from iris, retina and cornea in lentectomised eyes of *Xenopus laevis. Anat. Record* **145,** 214–215.

CAMPBELL, J. C. (1965). An immuno-fluorescent study of lens regeneration in larval *Xenopus laevis. J. Embryol. Exptl. Morphol.* **13,** 171–179.

COULOMBRE, A. J. (1956). The role of intraocular pressure in the development of the chick eye. I. Control of eye size. *J. Exptl. Zool.* **133,** 211–225.

COULOMBRE, A. J. (1957). The role of intraocular pressure in the development of the chick eye. II. Control of corneal size. *Arch. Ophthalmol. Chicago* **57,** 250–253.

COULOMBRE, A. J., and COULOMBRE, J. L. (1958). The role of intraocular pressure in the development of the chick eye. IV. Corneal curvature. *Arch. Ophthalmol. Chicago* **59,** 502–506.

DORRIS, F. (1938). Differentiation of the chick eye *in vitro. J. Exptl. Zool.* **78,** 385–416.

FISCHEL, A. (1903). Weitere Mittheilungen über die Regeneration der Linse. *Arch. Entwicklungsmech. Organ.* **15,** 1–138.

FREEMAN, G. (1963). Lens regeneration from the cornea in *Xenopus laevis. J. Exptl. Zool.* **154,** 39–65.

FREEMAN, G., and OVERTON, J. (1961). Lens regeneration from the cornea in *Xenopus laevis. Am. Zoologist* **1,** 448.

FREEMAN, G., and OVERTON, J. (1962). The effect of thyroxin on the competence for lens regeneration in *Xenopus. Anat. Record* **142,** 305.

IKEDA, Y. (1936). Beiträge zur Frage der Fähigskeit zur Linsenregeneration bei einer Art von *Hynobius* (*Hynobius unnangso,* Tago). *Arb. Anat. Inst. Sendai* **18,** 17–50.

IKEDA, Y. (1939). Zur Frage der Linsenpotenz der Hornhaut in spatembryonalen und larvalen Stadien bei einer Art von *Hynobius* (*Hynobius unnangso,* Tago). *Arb. Anat. Inst. Sendai.* **22,** 27–52.

JONES, K. W., and ELSDALE, T. R. (1963). The culture of small aggregates of amphibian embryonic cells *in vitro. J. Embryol. Exptl. Morphol.* **11,** 135–154.

KABAT, E. A., and MAYER, M. M. (1961). "Experimental Immuno-chemistry." Thomas, Springfield, Illinois.

LEWIS, W. H. (1905). Experimental studies on the development of the eye in amphibia. II. On the cornea. *J. Exptl. Zool.* **2,** 431–447.

MOSCONA, A. (1957). Formation of lentoids by dissociated retinal cells of the chick embryo. *Science* **125,** 598–599.

NIEUWKOOP, P. D., and FABER, J. (1956). "Normal table of *Xenopus laevis* (Daudin)." North Holland Publ., Amsterdam.

OVERTON, J., and FREEMAN, G. (1960). Lens regeneration in *Xenopus laevis*. *Anat. Record* **137,** 386.

REYER, R. W. (1954). Regeneration of the lens in the amphibian eye. *Quart. Rev. Biol.* **29,** 1–46.

REYER, R. W. (1956). Lens regeneration from homoplastic and heteroplastic implants of dorsal iris into the eye chamber of *Triturus viridescens* and *Amblystoma punctatum*. *J. Exptl. Zool.* **133,** 145–183.

SATO, T. (1951). Über die linsenbildende Fähigkeit des Pigmentepithels bei *Diemyctylus pyrrhogaster*. *Embryologia* **1,** 21–57.

SELMAN, G. G., and PAWSEY, G. J. (1965). The utilisation of yolk platelets by tissues of *Xenopus* embryos studied by a safranin staining method. *J. Embryol. Exptl. Morphol.* **14,** 191–212.

STONE, L. S. (1955). Regeneration of the iris and lens from retina pigment cells in adult newt eyes. *J. Exptl. Zool.* **129,** 505–534.

STONE, L. S. (1958). Lens regeneration in adult newt eyes related to retina pigment cells and the neural retina factor. *J. Exptl. Zool.* **139,** 69–84.

STONE, L. S. (1960). Regeneration of the lens, iris and neural retina in a vertebrate eye. *Yale J. Biol. Med.* **32,** 464–473.

STONE, L. S., and GALLAGHER, S. B. (1958). Lens regeneration restored to iris membranes when grafted to neural retina environment after cultivation *in vitro*. *J. Exptl. Zool.* **139,** 247–262.

TAKANO, K. (1958). On the lens effect in the Wolffian lens regeneration in *Triturus pyrrhogaster*. *Mie Med. J.* **8,** 385–403.

TÖRÖ, E. (1932). Die Regeneration der Linse in fruhen Entwicklungsstadien bei *Amblystoma mexicanum*. *Arch. Entwicklungsmech. Organ.* **126,** 185–206.

DNA Synthesis and the Incorporation of Labeled Iris Cells into the Lens during Lens Regeneration in Adult Newts[1]

RANDALL W. REYER

INTRODUCTION

Lens regeneration from the pigmented epithelium of the iris and neural retina regeneration from the pigmented retinal epithelium in newts are two clear cases of the metaplasia of a differentiated tissue into another kind of tissue. The experimental analysis of the tissue changes and interactions have been reviewed by Stone (1959, 1960, 1965), Reyer (1954, 1962), and Scheib (1965). Recently, the molecular events in the pigmented iris epithelium have been carefully studied by Yamada and co-workers (reviewed by Yamada, 1966, 1967a,b). Dedifferentiation and redifferentiation at the ultrastructural level have been described by Eguchi (1963, 1964), Karasaki (1964), and Dumont *et al.* (1970). Eisenberg and Yamada (1966), Eisenberg Zalik and Yamada (1967), and Yamada and Roesel (1969) studied DNA synthesis, movement of the labeled cells, and the cell cycle during lens regeneration from the iris.

The objectives of the research to be reported in this paper were: (1) to determine when DNA synthesis is first initiated in lens regeneration and which cells are active in DNA synthesis during the course of the regenerative process; (2) to correlate cessation of DNA synthesis with the onset of lens fiber differentiation in a closely staged series of lens regenerates; (3) to follow the populations of cells, which were labeled at frequent intervals during 20 days of regeneration, through subsequent stages of regeneration into a differentiated lens detached from the iris; (4) to determine, in this species, the duration of the period of time after injection during which thymidine-^{3}H is available in

[1] This investigation was supported by Research Grant NB 01544 from the National Institute of Neurological Diseases and Blindness of the National Institutes of Health and an Institutional Research Grant from The American Cancer Society.

the eye, and whether there is any additional recycling of isotope during the 2 weeks following injection; (5) to observe the effect of the extirpation of neural retina and lens on the initiation and pattern of DNA synthesis in both the pigmented epithelium of the iris and the pigmented, retinal epithelium during concurrent lens and retina regeneration. Some of this work has been reported in abstract form (Reyer, 1966, 1967). In general, the results of these experiments support the published results of Eisenberg and Yamada and provide an independent confirmation of their work. Furthermore, there are some differences in the interpretation of data on the cessation of DNA synthesis. A more detailed study than heretofore has been made, involving 103 cases, which traces the labeled cell populations during subsequent stages of lens differentiation; the results have been summarized in a series of diagrams and photographs. The studies of concurrent lens and retina regeneration were not undertaken by Eisenberg and Yamada.

MATERIALS AND METHODS

All experiments were performed on adult *Notophthalmus* (*Triturus*, *Diemictylus*) *viridescens viridescens*, collected in Massachusetts or West Virginia. Before operation the animals were anesthetized in 0.1% chloretone (Parke Davis) in distilled water and then immersed in a petri dish in amphibian Ringer's solution buffered with Tris-HCl to a pH of 7.0–7.2. The lens was removed from both eyes with pointed watchmaker's forceps through a transverse incision in the cornea (series 1–3). In the last group of experiments (series 4), both the lens and the neural retina were excised after an incision had been made around the dorsal one-half of the eye adjacent to the corneoscleral junction and that portion of the eyeball comprising the cornea, iris, and lens had been reflected laterally (Stone, 1950). Frequently, patches of pigmented retina could be seen still adhering to the extirpated neural retina. In most eyes, when the neural retina was torn from its marginal attachment, the unpigmented inner epithelium of the ciliary retina together with a narrow remnant of neural retina next to the *ora serrata* was not removed. This remnant tended to be larger in the part of the eyeball ventral to the incision. Following the above procedures, the reflected half of the eyeball was returned to its normal position, the cut edges of sclera opposed to each other and the dorsal eyelid pulled down over the incision. Recovery took place overnight in a moist chamber at 10–12°C. The newts were cultured in dechlorinated tap water at room temperature and fed three times a week with chopped beef liver.

In order to study the synthesis of deoxyribonucleic acid (DNA) at different stages in lens regeneration and to follow the labeled cells during the subsequent growth and differentiation of the lens, a group of animals was injected intraperitoneally with thymidine-^{3}H at a particular time interval after lentectomy and then the individual recipients were fixed at different periods of time after injection of the isotope. As a control to test the time after injection during which thymidine-^{3}H is available for uptake by cells synthesizing DNA, 14- to 16-day regenerating lens vesicles were implanted through a corneal incision into host eyes immediately after lentectomy at various times before and after administration of the isotope. The detailed data for these time intervals will be presented with the results for each series.

The data on the tritiated thymidine employed are as follows: thymidine-^{3}H, with a specific activity of 1.9 Ci/mmole, and a concentration of 0.5 mCi/ml, obtained from Schwarz Bio-Research, Inc., was used for series 1 and 4. For series 2 and 3, the labeled compound was designated as thymidine-methyl-^{3}H. The specific activity used for series 2 was also 1.9 Ci/mmole, and for series 3 and the control series, it was 14.0 Ci/mmole and 10.0 Ci/mmole, respectively. The concentrated isotope solution was diluted with sterile, distilled water to give the following dosages per animal: Series 1—a total of 3.75 μCi in 3 injections of 1.25 μCi each at hourly intervals. Series 2—a total of 6.00 μCi in 3 injections of 2.00 μCi each at hourly intervals. *Series 3*—a total of 4.00 μCi in a single injection. *Series 4*—a total of 3.75 or 5.00 μCi in 3 or 4 injections of 1.25 μCi each at hourly intervals. Control Series—1.50 μCi per gram body weight in a single injection. The animals used were of medium size in the range of 2.0–5 gm body weight. The largest animals were used in series 2 and the controls. Data from Hay and Fischman (1961) and Yamada and Roesel (1968), using two different methods, indicate that, following an intraperitoneal injection, thymidine-^{3}H remains available in the adult newt for no longer than 3 hours. The control experiments, reported here, confirm these data except for one exceptional case. The duration of isotope availability in the experiments described here thus ranges from a maximum of 6 hours (3+3) for the series with 4 injections to 3 hours or less for the series with a single injection.

Whole heads were fixed in Bouin's fluid, decalcified in 5% nitric acid in 80% ethanol and neutralized with 80% ethanol saturated with magnesium carbonate. The eyes and adjacent orbital bone were dehydrated in graded ethanols up to 95%, cleared in terpineol, rinsed in benzene, and embedded in paraplast (M. P. 56°-58°C). Serial cross sections, 10 μ in thickness, were cut parallel to the axis of the eye and mounted on clean glass slides using a gelatin-water solution (50 mg/25 ml) for floating and stretching the ribbons. The slides were dried and radioautographs prepared according to the methods of Hay and Fischman (1961) and Kopriwa and Leblond (1962). In series 1-4, the paraffin was removed in xylene, followed by 100% ethanol, ethyl ether, and drying in air. The slides were dipped, 5 at a time, in Kodak NTB-2 liquid emulsion at 40°-45°C using Lipshaw Peel-A-Way plastic slide grips. They were dried for 1.5-3 hours in an incubator at 28°C and 80% relative humidity and then placed in black plastic slide boxes with 20 gm of Drierite at 3°-4°C for an exposure of 3-10 weeks. Development was in Dektol 1:2 for 2 minutes at 17°-18°C. After development, fixation, and washing, the slides were stained through the emulsion with Ehrlich's acid hematoxylin and erythrosin, then dehydrated and covered in the usual fashion.

In order to obtain a lower background, the above procedure was modified in the control series according to the suggestions of Rose and Rose (1965). The eye was dissected from the bony orbit so that decalcification was not necessary, and serial sections, 8 μ in thickness, were coated with Kodak NTB-3 emulsion and exposed for 4-4.5 weeks at a temperature of 3°-4°C. These radioautographs were developed in Kodak D-11 developer for 4 minutes at 13°-14°C. After these procedures, a lower background was obtained than in the radioautographs previously described.

In the determination of the presence or absence of label over a particular nucleus, the number of silver grains over this nucleus was compared with the average number of grains over nuclei in adjacent unlabeled tissue. In any given slide, the intensity of labeling in a proliferating tissue could be determined by examination of the epidermis, oral mucosa, or corneal epithelium. The number of silver grains in the background showed some unevenness in many preparations, tending to be higher over cavities immediately adjacent to a tissue, as within the anterior or vitreous chambers. This was probably due to greater thickness of the emulsion at these sites. When very intense labeling of numerous, adjacent nuclei occurred, after a long exposure, there was considerable scattering of silver grains into the area next to the heavily labeled cells.

RESULTS

Initiation of DNA Synthesis

In order to determine the minimal time between lens removal and the beginning of DNA synthesis in the iris epithelium, thymidine-^3H was injected into groups of 6 animals in series 1 and 2 at 1–5 days after lens removal. The cases in each group were fixed at 3–5 hours and at 2, 5, 10, 14–16, and 20–21 or 25–26 days after injection (Table 1, A-C; Table 2). A total of 47 animals and 91 eyes were studied.

Except for an occasional cell in the connective tissue stroma, no labeled nuclei were found in the iris of the intact, normal eye of the adult newt. However, detection of silver grains is difficult over the iris epithelium because its nuclei are obscured by very heavy pigmentation. Occasionally, a few labeled nuclei were found in the region of the *ora serrata* at the margin of the neural retina. After isotope injection during the first 3 days after lentectomy, only rarely was there labeling of a nucleus in the iris epithelium. Nearly all these nuclei, in animals fixed either immediately after injection or up to 20 days later, had no enhancement of silver grains above the background level. Because there was extensive depigmentation of the cells developing into a lens, a label was much easier to detect in older stages. One to three labeled nuclei were found in the entire dorsal region of the iris epithelium in 7

TABLE 1

TIME INTERVALS AFTER LENTECTOMY FOR INJECTION OF THYMIDINE-^3H AND TIMES AFTER INJECTION WHEN THE CASES WERE FIXED[a]

Series 1	Injection of thymidine-^3H, days after operation	Fixation time after injection of thymidine-^3H						
		Hours	Days					
		3	2	5	10	15–16	20–21	25–26
A	1	0–0	1–1	2–1	4–4	8–7	—	10–10
B	3	1–1	3–2	2–2	4–4	8–8	—	10–10
C	5	1–1	2–2	3–3	6–6	9–9	10–10	—
D	7	2–2	3–3	4–4	7–7	9–9	11–11	—
E	10	3–3	4–4	5–6	8–6	10–10	11	—
F	15	4–5	7–6	9–7	10–10	10–10	—	11–11
G	20	9–9	9–9	10–10	10	—	9–11	—

[a] The one or two numbers for each animal indicate the stage of lens regeneration in each eye at fixation.

TABLE 2

TIME INTERVALS AFTER LENTECTOMY FOR INJECTION OF THYMIDINE-^{3}H AND TIMES AFTER INJECTION WHEN THE CASES WERE FIXED[a]

Series 2	Injection of thymidine-^{3}H, days after operation	Fixation time after injection of thymidine-^{3}H					
		Hours	Days				
		3–5	2	5	10	14–15	20
A	1	0–0	1	1–2	4–4	7	–
B	2	1–1	1–1	2–2	3–3	7–7	10–9
C	3	1–1	1–1	2–2	4–4	9–9	10–10
D	4	1–1	1–1	2–2	5–5	6–6	10–10
E	5	1	1–1	2–2	5–5	9–9	10–10

[a] The one or two numbers for each animal indicate the stage of lens regeneration in each eye at fixation.

cases, injected 1 or 2 days after lentectomy, and 1 to 6 labeled nuclei in 7 cases injected 3 days after lens removal. With one exception, all the labeled nuclei were in specimens which had been fixed 2 or more days after injection of the isotope. Since it is known from the electron microscopic studies of Eguchi (1963) that macrophages become intimately associated with the iris epithelium during depigmentation, there is the possibility that some of the labeled nuclei belonged to these elements, probably recruited from the connective tissue or blood. In one of the cases, however, 3 labeled nuclei with grain counts of approximately 30, 40, and 50 were located in the primary lens fibers of a stage[2] 8 lens, injected 3 days after lens removal and fixed 15 days after injection (Fig. 1).

The earliest time at which DNA synthesis was observed in an appreciable number of iris cells was 4 days after lens removal. In 5 out of the 12 cases, there were scattered, labeled nuclei in the inner lamina of the dorsal iris at regeneration stages[2] 1 and 2. After administration of the isotope at this time, silver grains were aggregated over scattered nuclei at 5 hours after injection as well as in cases fixed after 2 and 5 days. Five days after lens removal, a larger number of nuclei in the inner lamina of the dorsal iris epithelium were labeled in 22 out of 23 cases. The location of most of the labeled nuclei extended from the pupillary margin one half the distance to the ciliary retina.

[2] The stages of regeneration refer to the morphological stages of Sato (1940) which have also been illustrated by Reyer (1954, 1962), Stone and Steinitz (1953), and Yamada (1967b).

A very few mitotic figures (metaphases or anaphases) appeared at regeneration stage 2, 6–8 days after lens removal. These increased in frequency among the depigmenting cells at regeneration stages 3 and 4. There appeared to be a 2-day interval between initiation of DNA synthesis and the first mitoses of the cells in the dorsal iris epithelium.

Subsequent Development of Labeled Iris Cells

The lens vesicles and lenses which regenerated from iris, labeled 5 days after lens extirpation at stage 1, had labeled nuclei in the inner wall of the lens vesicle, in the primary lens fibers derived from this area as well as in the dorsal secondary lens fibers and in the dorsal part of the lens epithelium derived from the inner iris lamina. Only occasionally were labeled nuclei observed in the outer lamina of the iris or in the outer lens epithelium. These results are summarized in the drawings of Fig. 2 and illustrated by photomicrographs in Figs. 3–5. There was a diluting of the label due to continued mitoses of the labeled cells in the cases fixed at regeneration stage 5 and later. Even in old regenerates, however, some cells in the inner lamina of the iris remained heavily labeled. It would appear that these cells synthesized DNA but then either failed to divide or else went through only one mitotic cycle so that marked dilution of the label did not occur.

When 7 days had elapsed between lens extirpation and injection of the isotope at regeneration stage 2 (Table 1, series 1, group D), then many labeled cells were observed in the inner lamina of the dorsal iris. These were most numerous near the pupil and decreased toward the ciliary retina. Furthermore, labeled cells were now present in the outer lamina of the iris adjacent to the pupil. Scattered, labeled nuclei were also found in the inner, unpigmented epithelium of the ciliary retina. The label could be followed into the entire lens vesicle and into the primary lens fibers and lens epithelium developing from this vesicle. When secondary lens fibers appeared, many of these were also labeled. Similar results, but with a further increase in the number of labeled nuclei, were obtained with thymidine-^{3}H injection 10 days after operation at regeneration stage 3 (series 1, group E). In both groups, the label in the lens epithelium gradually became diluted in older stages due to continued mitoses in this tissue (Figs. 6–9).

In series 1, group F, the eyes were exposed to thymidine-^{3}H 15 days after lens extirpation. The single case, fixed 3 hours after injection, had a lens regenerate of stage 4 in the right eye and of stage 5 in the left eye. There were numerous, uniformly distributed, labeled nuclei throughout these lens vesicles. After regeneration for different lengths

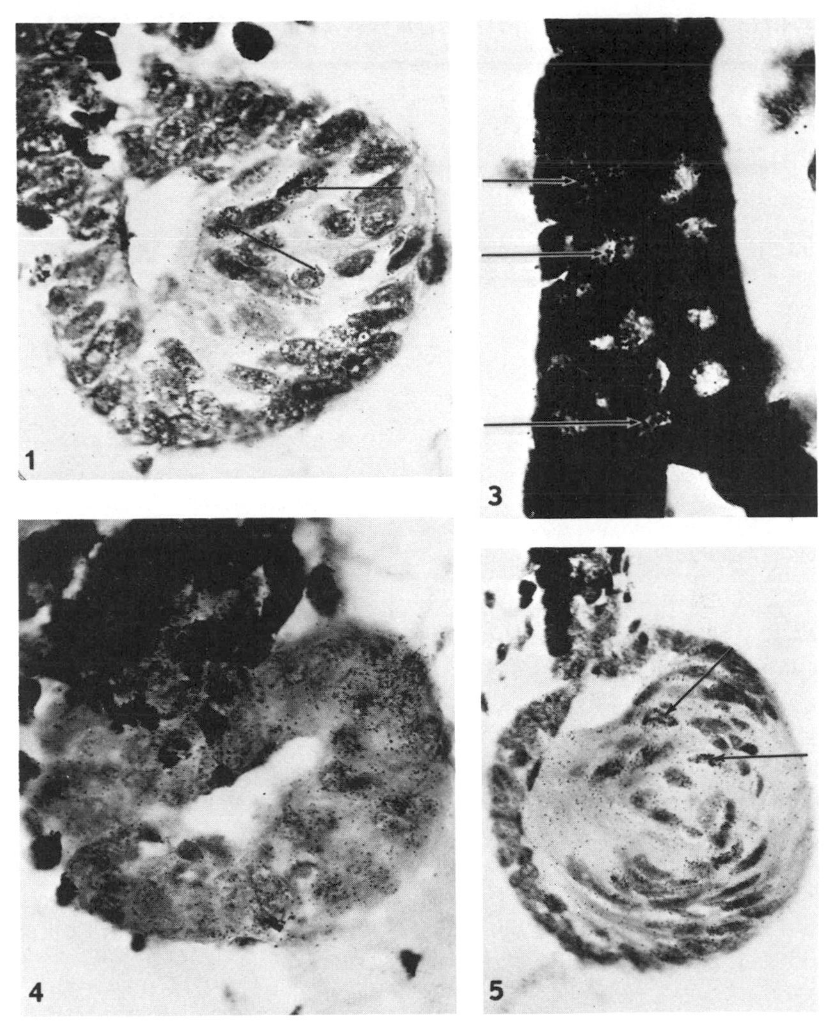

FIG. 1. Radioautograph of regenerating lens of stage 8 in right eye. TdR-^{3}H injected 3 days after lentectomy, fixed 15 days after injection. Nuclei of two lens fibers are labeled (indicated by arrows). × 305.

FIG. 3. Radioautograph of regenerating lens of stage 2 in left eye. TdR-^{3}H injected 5 days after lentectomy, fixed 2 days after injection. Labeled nuclei (indicated by arrows) in inner lamina of iris epithelium. × 775.

FIG. 4. Radioautograph of regenerating lens of stage 5 in right eye. TdR-^{3}H injected 5 days after lentectomy, fixed 10 days after injection. Labeled nuclei in margin of dorsal iris and inner part of lens vesicle. Perhaps an occasional cell with light label in outer part of lens vesicle. × 305.

THYMIDINE-H^3 INJECTED 5 DAYS AFTER LENS REMOVAL

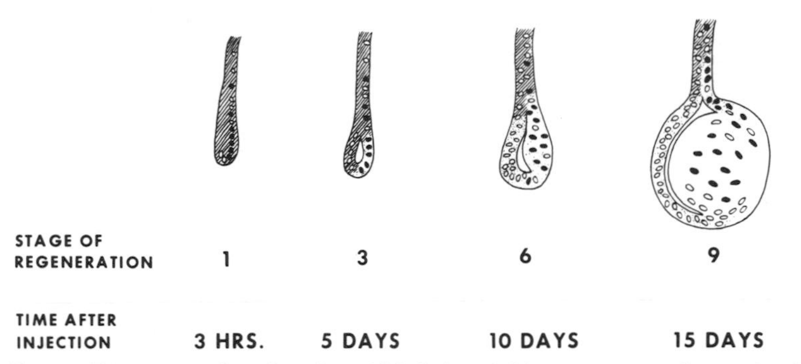

FIG. 2. Diagram to show location of labeled nuclei in regenerating lenses developing from the dorsal iris after injection of TdR-^{3}H 5 days after lentectomy. Clear nuclei unlabeled, black nuclei labeled; diagonally hatched—pigmented or partly depigmented epithelium; stippled—depigmented epithelium; blank—lens fibers.

of time after injection, label was observed in the lens epithelium and in some of the primary lens fibers as well as in the secondary lens fibers when they appeared (Figs. 10–14). However, there was a lower frequency of labeled primary lens fibers than was found when the regenerate was exposed to the isotope at stage 3. This varied between individual specimens as can be seen in the right and left eyes of an animal fixed 5 days after injection. The stage 7 regenerate in the left eye (Fig. 12) had numerous labeled primary lens fibers while the stage 9 regenerate in the right eye (Fig. 13) contained only a few labeled primary lens fibers. At the time of injection, the latter probably had been a little more advanced in differentiation than the former. Eighteen animals and 35 eyes were studied in series 1, groups D, E, and F.

Cessation of DNA Synthesis in Prospective Primary Lens Fibers

In order to obtain more exact data concerning the time at which DNA synthesis ceases in those cells starting to differentiate into lens fibers, another experiment (series 3) was undertaken in which a single injection of 4 μCi of thymidine-^{3}H was administered at daily intervals

---

FIG. 5. Radioautograph of regenerating lens of stage 9 in right eye. TdR-^{3}H injected 5 days after lentectomy, fixed 15 days after injection. Labeled nuclei (arrows) visible in primary and secondary lens fibers, lens epithelium now unlabeled. $\times 220$.

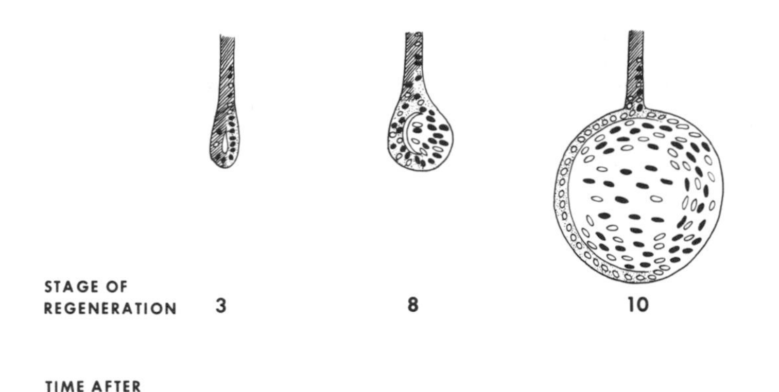

FIG. 6. Diagram to show location of labeled nuclei in regenerating lenses developing from the dorsal iris following injection of TdR-^{3}H 10 days after lentectomy.

from 6 to 20 days after lens removal. All animals were fixed 3–4 hours after injection. A total of 30 animals and 59 eyes were included in this series. The rate of lens regeneration was slightly faster than in series 1 so that the critical regeneration stages 4–6 were reached from 10–15 days after operation.

Staging of the regenerates is not as precise in adult newts as in the larval stages described by Reyer (1954). The regenerating lens vesicle of the adult is made up of a larger number of cells, the inner epithelium may be folded and the nuclei arranged somewhat irregularly. The nasotemporal extent of the depigmented region at the pupillary margin is also variable (Stone and Steinitz, 1953; Yamada, 1967b).

Typically, in stage 4, the lens vesicle was composed of cuboidal to low columnar cells. These were attached to the inner and outer iris laminae and projected into the pupil only a short distance beyond the still pigmented cells of the outer iris lamina. At stage 5, this vesicle had increased in size, the lumen was larger unless filled by folds of the vesicle epithelium and some of the cells in the inner wall of the vesicle were becoming more columnar. At stage 6, these latter cells had elongated further to form a little group of differentiating, primary lens fibers. When exposed to thymidine-^{3}H, many but not all the cells of the lens vesicle were labeled. In all cases at stage 4 and in some cases at stage 5, labeled and unlabeled cells were intermixed but the proportion of labeled to unlabeled cells was often greater in the outer wall of the vesicle than in the inner wall (Fig. 15). When several adjacent cells

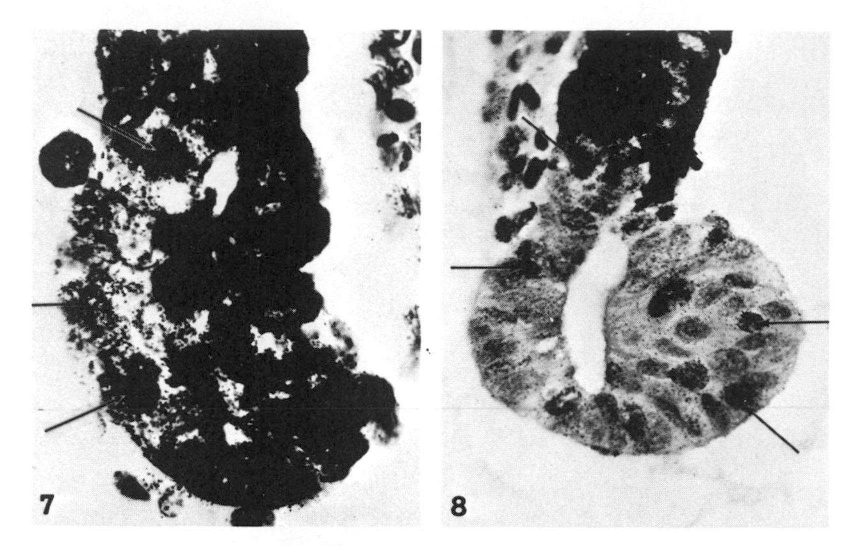

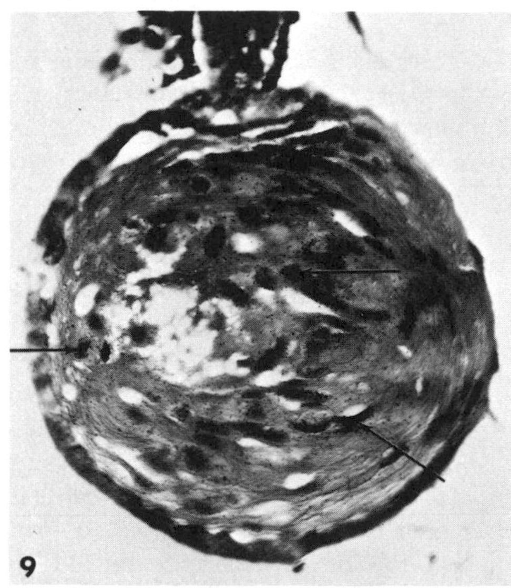

FIG. 7. Radioautograph of regenerating lens of stage 3 in left eye. TdR-^{3}H injected 10 days after lentectomy, fixed 3 hours after injection. Labeled nuclei in inner lamina of iris epithelium are indicated by arrows. The larger black granules are melanin. × 423.

FIG. 8. Radioautograph of regenerating lens of stage 8 in right eye. TdR-^{3}H injected 10 days after lentectomy, fixed 10 days after injection. Labeled nuclei (arrows) in stalk connecting lens with iris, lens epithelium, and primary lens fibers. × 220.

FIG. 9. Radioautograph of regenerating lens of stage 10 in right eye. TdR-^{3}H injected 10 days after lentectomy, fixed 16 days after injection. Labeled nuclei (arrows) in primary and secondary lens fibers. × 195.

THYMIDINE-H3 INJECTED 15 DAYS AFTER LENS REMOVAL

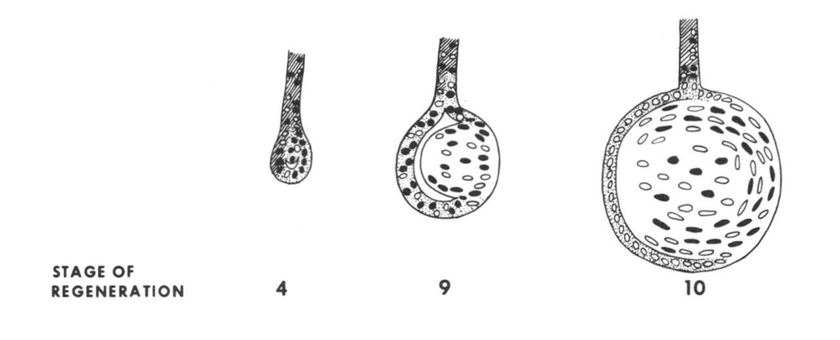

TIME AFTER INJECTION — 3 HRS. — 5 DAYS — 15 DAYS

FIG. 10. Diagram to show location of labeled nuclei in regenerating lenses developing from the dorsal iris following injection of TdR-^{3}H 15 days after lentectomy.

in the inner wall were not labeled, these unlabeled cells did not continue through more than three consecutive serial sections. Frequently at stage 5, however, a small group of low columnar, unlabeled cells was observed in the inner wall, extending through four or more serial sections (Figs. 16 and 17). At stage 6, the beginning lens fibers were uniformly unlabeled while the rest of the lens vesicle showed a random distribution of labeled cells (Fig. 18, Table 3). It can be concluded that synthesis of DNA stops in the prospective primary lens fiber cells which are just beginning to increase in height at regeneration stage 5. In older stages, differentiating primary and secondary lens fibers uniformly fail to label, while most of the cells of the lens epithelium and those connecting the regenerate to the iris itself are labeled.

Subsequent Development of the Labeled Lens Epithelium

When the epithelial cells, labeled at regeneration stage 9, 20 days after lentectomy, were traced in specimens fixed at intervals after injection of the isotope (series 1, group G), the label was absent from

---

FIG. 11. Radioautograph of regenerating lens of stage 4 in right eye. TdR-^{3}H injected 15 days after lentectomy, fixed 3 hours after injection. Densely labeled nuclei in depigmented and partly depigmented cells of lens vesicle. × 332.

FIG. 12. Radioautograph of regenerating lens of stage 7 in left eye. TdR-^{3}H injected 15 days after lentectomy, fixed 5 days after injection. Nuclei with a medium label in both lens epithelium and primary lens fibers. × 305.

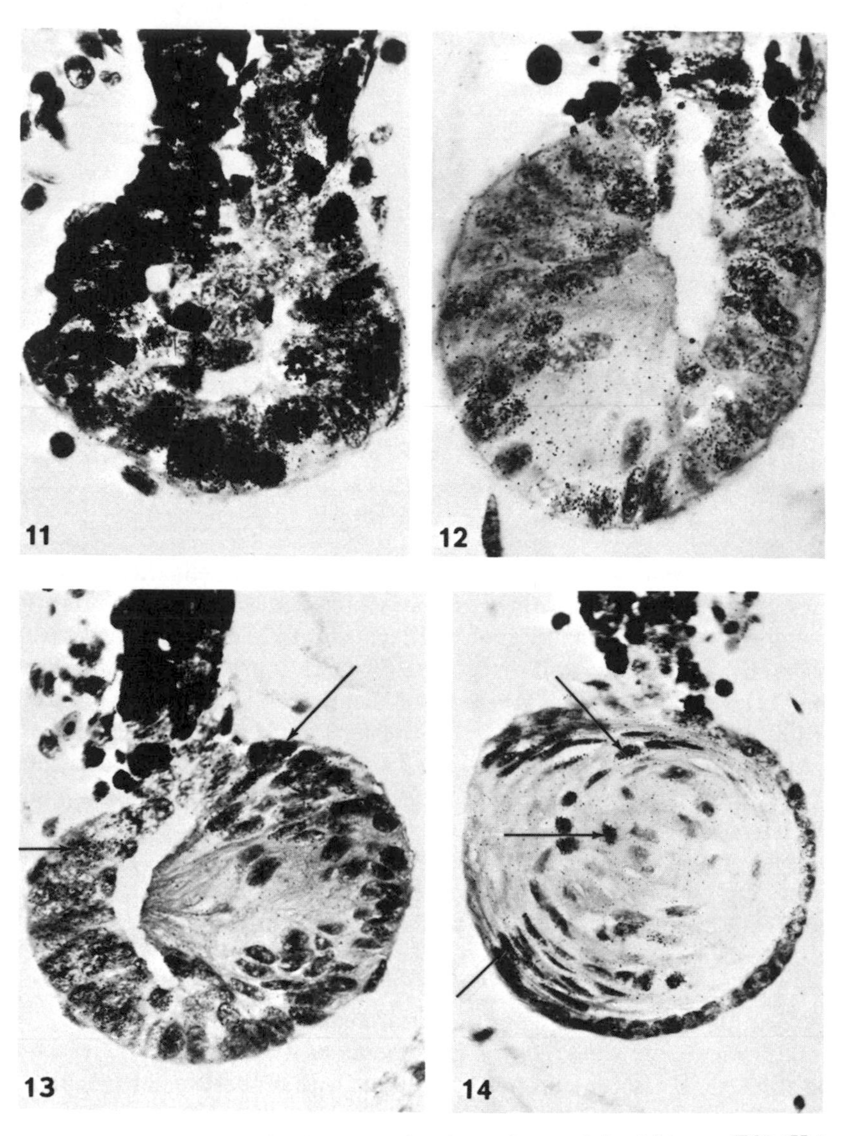

FIG. 13. Radioautograph of regenerating lens of stage 9 in right eye. TdR-^{3}H injected 15 days after lentectomy, fixed 5 days after injection. Labeled nuclei (arrows) in lens epithelium and secondary lens fibers. Very few primary lens fibers are labeled. Same animal as shown in Fig. 12. × 220.

FIG. 14. Radioautograph of regenerating lens of stage 10 in left eye. TdR-^{3}H injected 15 days after lentectomy, fixed 10 days after injection. Labeled nuclei (arrows) in a few of the primary lens fibers and in many secondary lens fibers. × 195.

TABLE 3
LENS REGENERATION STAGES WHEN A GROUP OF UNLABELED CELLS FIRST APPEARED IN THE INNER WALL OF THE LENS VESICLE

Series 3	Total No.	Number of eyes with group of unlabeled cells present	Number of eyes with group of unlabeled cells absent
Stage 4	10	0	10
Stage 5	8	4	4
Stage 6	4	4	0

the primary and first-formed secondary lens fibers. Many labeled cells were present, however, in the lens epithelium and in the outer secondary lens fibers derived from it (Figs. 19–22). This labeling remained heavy in the innermost of the labeled, secondary lens fibers but was gradually diluted by mitoses so that it was no longer detectable in the lens epithelium of the specimen fixed 21 days after injection. It is possible, therefore, to follow a group of secondary lens fibers from their origin in the lens epithelium through their differentiation in the equatorial growth zone and their gradual displacement into deeper layers of the lens as appositional growth continues.

Concurrent Lens and Neural Retina Regeneration

Both the lens and neural retina were excised in series 4. Thymidine-^{3}H was injected into 7 groups of animals at 4, 7, 10, 15, 19–20, 25, and 35 days after operation, and the cases in each group were fixed at 3 hours and 1, 2, 5, 10, 15–16, 20, and 30 days after injection. Radioautographs of serial sections were then prepared. The data on injection and fixation times and on the stage of lens regeneration are summarized in Table 4. A comparison between the rate of lens development after removal of the lens alone and after removal of both the lens and neural retina is shown in Fig. 23. There was a delay of 10–15 days in depigmentation and in the beginning of lens fiber formation in the latter situation. By 40 days after operation, large lenses of stages 11 or 12 were present in the eyes of both groups. As previously observed by Stone (1950), Hasegawa (1958), and Stone and Steinitz (1953), there was a concurrent regeneration of both neural retina and lens. The average delay in the development of the lens was somewhat larger here than the approximately 5 day lag depicted in Fig. 1 of Stone and Steinitz. The source of the regenerated neural retina appeared to be 2-fold; it formed in part from

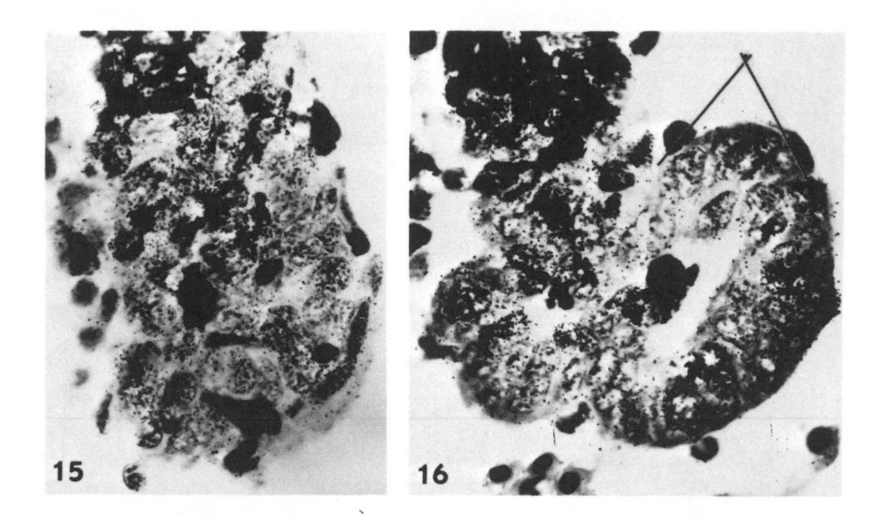

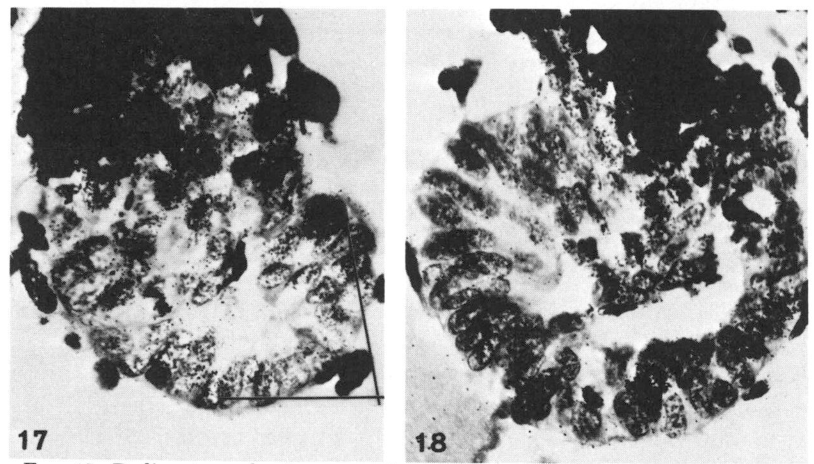

FIG. 15. Radioautograph of regenerating lens of stage 4 in right eye. TdR-^{3}H injected 11 days after lentectomy, fixed 3 hours after injection. Numerous labeled nuclei distributed among the depigmented cells of the lens vesicle. × 332.

FIG. 16. Radioautograph of regenerating lens of stage 5 in right eye. TdR-^{3}H injected 14 days after lentectomy, fixed 3 hours after injection. Labeled nuclei in both inner and outer walls of the lens vesicle. However, a small group of unlabeled nuclei was present in the dorsal portion of the inner wall (between intersecting lines) which extended over 5 adjacent sections. ×332.

FIG. 17. Radioautograph of regenerating lens of stage 5 in right eye. TdR-^{3}H injected 13 days after lentectomy, fixed 3 hours after injection. Labeled nuclei in both inner and outer walls of the lens vesicle. However, a larger group of unlabeled nuclei was located in the ventral part of the inner wall (between intersecting lines) which extended over 8 adjacent sections. ×332.

FIG. 18. Radioautograph of regenerating lens of stage 6 in left eye. TdR-^{3}H injected 15 days after lentectomy, fixed 3 hours after injection. Labeled nuclei were restricted to the outer and ventral wall of the lens vesicle. The elongating primary lens fibers in the inner wall were unlabeled. × 276.

any of the inner, unpigmented retina remaining adjacent to the *ora serrata*, which was actively synthesizing DNA by 4 days after operation, and in part from the pigmented retinal epithelium where initiation of DNA synthesis was somewhat delayed.

The initiation of DNA synthesis in the dorsal iris epithelium was also slightly delayed when the neural retina as well as the lens was removed. Labeled nuclei were extremely rare up to 4 days after opera-

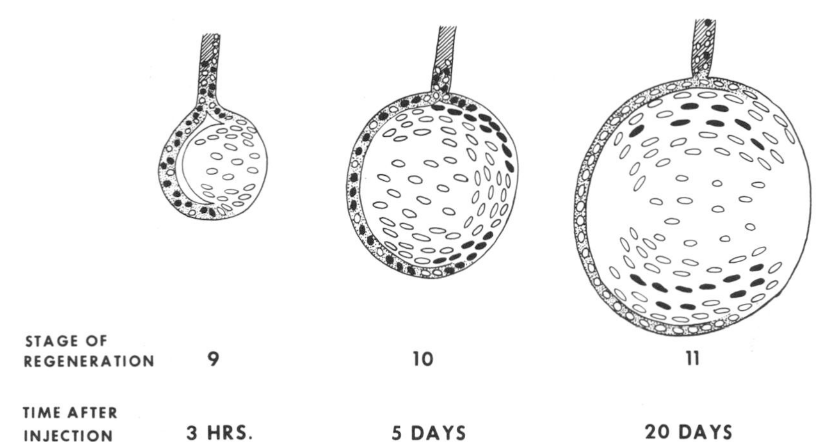

FIG. 19. Diagram to show location of labeled nuclei in regenerating lenses developing from the dorsal iris following injection of TdR-^{3}H 20 days after lentectomy. At the time of isotope injection, only nuclei in the lens epithelium and the stalk of the lens were labeled. Labeled cells could subsequently be traced into the secondary lens fibers.

FIG. 20. Radioautograph of regenerating lens of stage 9 in right eye. TdR-^{3}H injected 20 days after lentectomy, fixed 3 hours after injection. Labeled nuclei in stalk of lens and in middle of lens epithelium. No label was present over nuclei of primary or secondary lens fibers. × 220.

FIG. 21. Radioautograph of regenerating lens of stage 10 in left eye. TdR-^{3}H injected 20 days after lentectomy, fixed 5 days after injection. Labeled nuclei visible in lens epithelium, lens stalk, and outermost secondary lens fibers (arrows). × 195.

FIG. 22. Radioautograph of regenerating lens of stage 10½ in left eye. TdR-^{3}H injected 20 days after lentectomy, fixed 10 days after injection. Label restricted to outer layers of secondary lens fibers (arrows). × 166.

FIG. 24. Radioautograph of stage 2 of lens regeneration in right eye. TdR-^{3}H injected 7 days after removal of lens and neural retina, fixed 1 day after injection. Labeled nuclei present in both inner and outer laminae of iris epithelium (arrows). The other dark granules are melanin pigment. × 775.

FIG. 25. Radioautograph of regenerating lens of stage 4 in right eye. TdR-^{3}H injected 25 days after removal of lens and neural retina, fixed 3 hours after injection. Heavily labeled nuclei in many of the depigmented, lens vesicle cells. × 332.

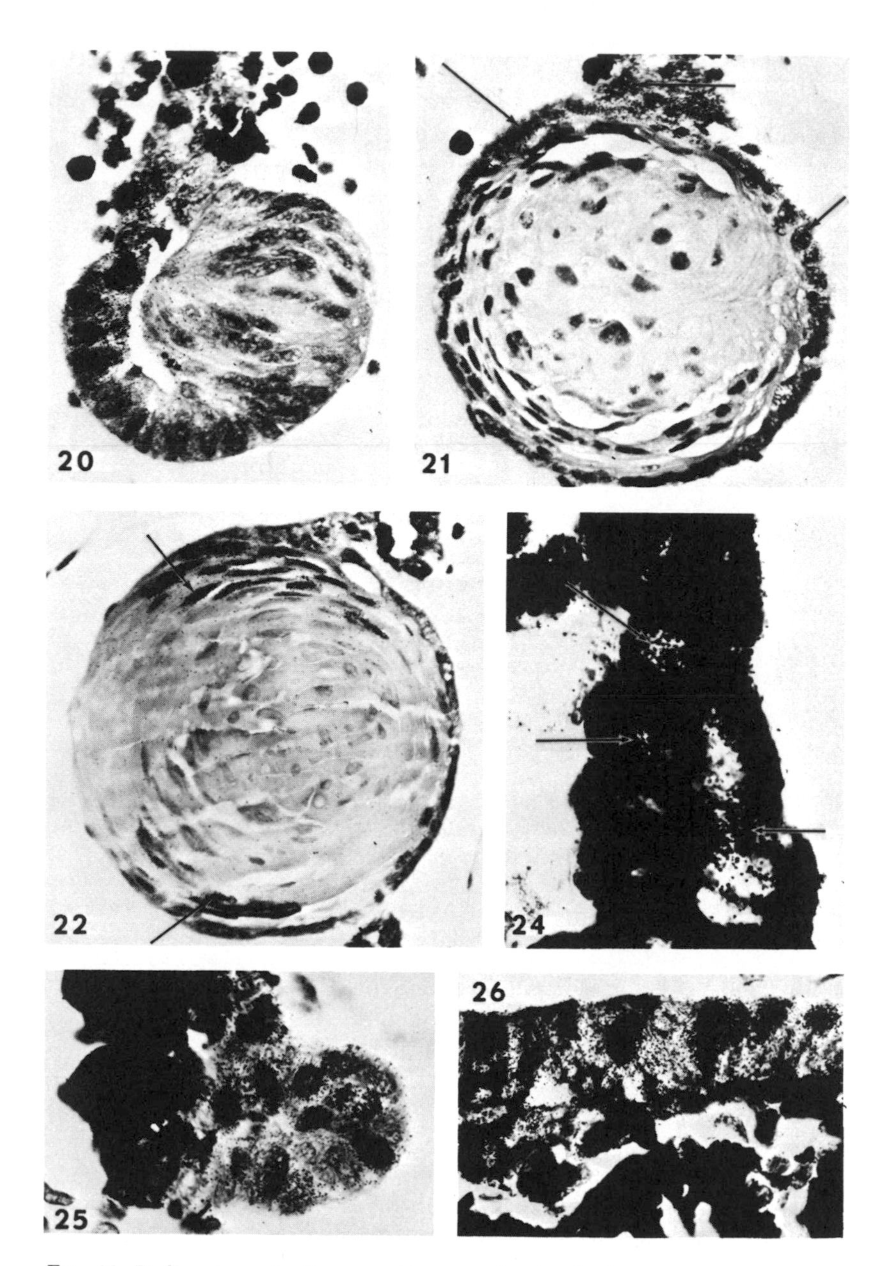

FIG. 26. Radioautograph of regenerating neural retina in the same eye as illustrated in Fig. 25. A single layer of depigmented cells has formed, many of which have a dense label over their nuclei. $\times 332$.

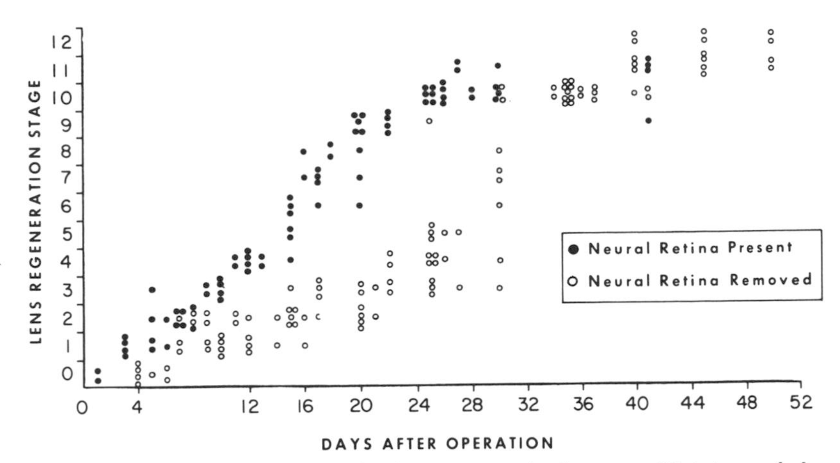

FIG. 23. Chart comparing the rate of lens regeneration in terms of Sato's morphological stages in eyes with intact neural retina and in eyes where lens and neural retina were removed at the same time. Each solid or clear circle represents one regenerating lens. Symbols for lenses at the same stage, fixed on the same day, are included between the lines above and below the number for that stage.

tion, but they appeared regularly in the inner lamina of the iris by 7 days after operation and could be traced into the lens vesicle in cases fixed at longer periods after injection. This was at the same time that labeled nuclei were first observed in the pigmented retina and indicates that DNA synthesis was initiated simultaneously in these two pigmented epithelia (Fig. 24). Scattered, labeled nuclei continued to be found in the dorsal iris of animals injected 10, 15, and 20 days after operation. These labeled iris cells could now be traced into primary and secondary lens fibers when fixed 15 or more days after exposure to the isotope. However, during this period, labeling was more extensive in the cells of the pigmented retina which were beginning to enlarge and undergo depigmentation. In some of the more advanced cases, depigmentation (stage 3) had also begun in the dorsal iris.

When a lens vesicle (stage 4) appeared by 25 days after operation, many of the nuclei of the depigmented cells were labeled. At this time, the regenerating neural retina consisted of one to several layers of unpigmented cells with many labeled nuclei (Figs. 25 and 26). Mitoses in the iris epithelium were first observed in one animal, fixed 15 days after operation, where a single mitotic figure was detected in the inner lamina of the dorsal iris epithelium of each eye at regeneration stage 2.

TABLE 4

Time Intervals after Lens and Neural Retina Extirpation for Injection of Thymidine-^{3}H and Times after Injection When the Cases Were Fixed

	Injection of thymidine-^{3}H, days after operation	Fixation time after injection of thymidine-^{3}H							
		Hours	Days						
		3	1	2	5	10	15–16	20	30
A	4	0–0 0–0	0	0–0	2–2	2–1	—	—	—
B	7	1–1 2	2–2	1–1	2–1	2–3	3–3	5–3	10–10
C	10	1–1 1–1	2–2	1–1	2–2	3–3	5–5	10–10	10–11
D	15	2–2 3–2	1–2	3–3	2–2	4–5	7–8	10–10	12–12
E	19–20	2–2	3–2	4–4	3–3	4–3	10–10	11–11	12–12
F	25	3 9 4 4–4	5–4	—	6 7	10 10–10	10–10	11–11	12
G	35	10–10 10–10	10–10	10	12–12	11–11	11–11	12–12	13–13 13–13

[a] The one or two numbers for each animal indicate the stage of lens regeneration in each eye at fixation.

Subsequently no more than one or two mitotic figures were found until 25 days after operation. There was one case, which had reached regeneration stage 5 at this time, in which numerous mitotic figures could be observed among the cells of the lens vesicle. These results indicate that DNA synthesis in the dorsal iris can be initiated in the absence of nearly all the neural retina. Based on the apparent absence of mitotic figures during the first 15 days after operation, there appears to be at least a 5- to 8-day and probably longer delay in the onset of mitosis for those cells which are activated to DNA synthesis at 7 and 10 days after operation. They must therefore remain in a prolonged S or G_2 phase before proceeding to M. This delay was correlated with retarded depigmentation of those cells destined to form the lens vesicle. Both active mitosis and depigmentation seem to require a continuous layer of unpigmented neural retina.

Control Series

In order to trace accurately a population of labeled cells over a period of time in a series of animals fixed at different time intervals after exposure to an isotope, it is essential that no cells become labeled following the initial period when the isotope was available. Therefore, an implantation experiment was performed in order to obtain information on the time after injection during which thymidine-^{3}H is available for uptake by cells, located in the eye, which are synthesizing DNA. A total of 33 donors and 33 hosts, both males and females, were used. The lenses were removed from both eyes of the donors and lens regeneration allowed to proceed for 14–16 days. The host lens was extirpated and a sector of mid-dorsal iris, including the regenerating lens vesicle, was removed from a donor eye and inserted through the corneal incision into the pupil of each of the host eyes. The host was allowed to recover in a moist chamber for 24 hours or until fixation. Injections of thymidine-^{3}H and operations were performed according to the following schedule. In group A (8 eyes), the hosts were injected 5 days, 1 day and 2 hours after operation and fixed in Bouin's fluid 2 hours after injection. In group B (22 eyes), the hosts were injected before the operation and then the lens was removed and the implant inserted into the eye chamber at 40, 80, and 130 minutes, 3.5, 5, 6.5, 8.5, 10.5, 12.5, 16, and 20 hours after injection. Fixation was 2 hours after operation. In group C (36 eyes), the hosts were again injected beforehand and the operation was performed 1, 2, 3, 4, 6, 8, 10, 12, and 14 days after injection. Fixation was 1 day after operation for the 1- to 3-day groups and 2 days after operation for the 4- to 14-day groups. Radioautographs were prepared as before, but with the modifications recommended by Rose and Rose (1965).

The depigmented lens vesicle cells of the implant were studied for evidence of uptake of thymidine-^{3}H. In three cases in group A (6 eyes), where the implant was already present in the eye at the time of isotope injection, numerous heavily labeled cells were found in the lens vesicle. These results indicate that the injected isotope is accessible to an implant which is lying free within the pupil and vitreous chamber. When implants were made at 40 and 80 minutes after isotope injection, the lens vesicle cells were also labeled, but with a decreased grain count (average of 53 grains per nucleus). In the specimen where the implant was made 130 minutes after injection, nearly all the depigmented cells of the lens vesicle were unlabeled except for a few where a label was barely discernible (8–12 grains per nucleus). With one exception, no labeled cells were found in the implanted lens vesicles in the other cases, where the time elapsed between isotope injection and implantation ranged from 3.5 hours to 14 days. These data would appear to indicate that a population of cells is labeled over a period between 2 and 3 hours by a single injection of thymidine-^{3}H at a dose level up to 1.5 μCi per gram body weight. Usually, no labeling of additional cells occurs later on due to subsequent access to the isotope. In some of the cases, fixed 8 or more days after injection, cells with a medium to heavy label were scattered in the vicinity of the implant. Occasionally, one was found within the depigmented epithelium itself. These were interpreted as cells of host origin, probably macrophages, which had become intimately associated with the depigmenting iris cells as discussed earlier. These results confirm those obtained by Yamada and Roesel (1968) in similar experiments carried out over the first 8 hours after injection of thymidine-^{3}H and extend the period of time surveyed to 14 days after administration of the isotope.

One exceptional case raised the possibility of occasional recycling of thymidine-^{3}H. In this specimen, the implant was made 2 days after injection followed by fixation 1 day later. Numerous cells in the implanted lens vesicles of both eyes were lightly labeled (average grain count 13 grains per nucleus). Further study will be required to determine whether this type of result occurs with any frequency.

DISCUSSION

Yamada and Roesel (1969) made a detailed study of the beginning of DNA synthesis during early lens regeneration by counting the labeled and unlabeled cells in radioautographs of paraffin sections only 3 μ in thickness. In their preparations, no iris epithelial cells were labeled during the first 2 days after lentectomy; an occasional labeled cell was detected 3 days after operation while 6.8% of the total cell population in the dorsal iris was labeled 4 days after lens removal and 25% after 5 days. The experiments reported here, which used the technique of fixation over a period of 3 hours to 20 days after injection, gave the same results and demonstrate that DNA synthesis is initiated 3–4 days after lentectomy. The single, positive case observed after injection at 3 days after lens removal and fixation 15 days later provides less critical evidence than the series of specimens with labeled nuclei which were exposed to the isotope 4 and 5 days after operation and preserved from 3 hours to many days after injection. Since rRNA synthesis begins 2 days after lens removal (Reese *et al.*, 1969), it seems to be clear that it precedes DNA synthesis by 1 or 2 days during the initial stages of lens regeneration.

A second important event in lens regeneration is the cessation of DNA synthesis in those cells beginning to differentiate into lens fibers. Eisenberg and Yamada (1966) and Yamada (1966, 1967b) have emphasized that this is the initial event in lens fiber differentiation preceding, in the primary lens fiber line, the beginning of synthesis of either alpha and beta or gamma crystallins. In their work, a small group of unlabeled nuclei was found in the inner part of a lens vesicle when thymidine-^{3}H was injected 11 days after lentectomy and the eye was fixed 12 hours after injection at stage 4. When similar specimens were fixed at 4 or 9 days after injection, both labeled and unlabeled cells were found among the developing primary lens fibers. When the injections were made at stages 6–8, 15 days after lentectomy, then all the primary lens fibers failed to label.

The work reported here analyzes this problem on a larger number of more closely staged specimens and is in general agreement with the conclusions of Eisenberg and Yamada. However, there appears to be one small point of disagreement, namely, that clear evidence for the permanent arrest of DNA replication in prospective primary lens fibers first appears at regeneration stage 5 rather than at regeneration stage 4 as proposed by Eisenberg and Yamada. This conclusion is based on the following considerations. Even in a tissue in which all cells are synthesizing DNA and dividing, not all of these cells would be labeled following a 3- to 5-hour exposure to thymidine-^{3}H when the DNA-S phase occupies 19 hours in a cell generation time of 65 hours as estimated by Eisenberg Zalik and Yamada (1967). Therefore, it cannot be assumed that scattered, unlabeled cells have permanently ceased DNA synthesis. In order to conclude that DNA synthesis has come to a permanent halt in a certain population of cells, there should be a group of adjacent, unlabeled cells appearing in several consecutive sections of the series. Further evidence would consist in the group of cells in this location remaining unlabeled in a closely staged series of increasingly older regenerates. Such a group of unlabeled cells was not observed at regeneration stage 4 and appeared for the first time among the cells which were beginning to elongate in the inner lamina of the lens vesicle at stage 5. By stage 6, these included all of the cells beginning to differentiate into primary lens fibers.

When the lens epithelium only was labeled, as in all cases injected after some lens fibers had differentiated, and the specimens were fixed at different time intervals after lentectomy, then the laying down of labeled, secondary lens fibers in concentric layers on the surface of the unlabeled, primary lens fibers could be followed. These results support similar findings of Eisenberg and Yamada (1966) and were, in addition, carried to somewhat older lenses.

Since the experiments of Stone and Steinitz (1953), it has been established that the simultaneous removal of neural retina and lens from the eye of the adult newt results in neural retinal regeneration followed by delayed lens regeneration. In their experiments, there was considerable variation in the time at which the regenerated lens attained a particular stage both in eyes with intact neural retina and in eyes with the neural retina removed. The delay in lens regeneration in the latter group, through the stages of depigmentation and beginning fiber formation, was approximately 5–8 days. A somewhat longer delay of 10–15 days was observed in the experiments reported here.

Of interest is the observation that DNA synthesis occurs in the cells of the pigmented epithelium of both iris and retina by 7 days after lens and neural retina extirpation. Subsequently, DNA synthesis, depigmentation and cell enlargement proceed more rapidly in the pigmented retina. Both depigmentation and mitotic activity appear to be markedly delayed in the pigmented iris epithelium when the neural retina has been removed and is in the early stages of reconstitution. The cells most active in DNA synthesis and cell division during these early stages of neural retina regeneration are the remnants of the inner, unpigmented neural retina immediately adjacent to the *ora serrata.* In the normal eye of the adult newt, only an occasional cell in this region shows DNA synthetic activity while none is observed in the remainder of the neural and all the pigmented retina. Extirpation of the lens and neural retina therefore activates DNA synthesis first in this region followed within a week by similar activity in the pigmented epithelium of dorsal iris and retina. The neural retina is reconstituted by contributions from both of these two sources. That developing from the marginal neural retina is always farther advanced in differentiation that that developing from the pigmented retina. As soon as a thin layer of depigmented, prospective neural retinal cells have formed, then lens regeneration from the iris proceeds in a normal fashion. Hasegawa (1958) concluded that neural retina regenerated from both the pigmented and ciliary, retinal epithelium following lens and neural retina extirpation. In a second paper (1965), the role of the marginal retina was further established by observing reconstitution of a new neural retina after surgical ablation of the entire neural and pigmented retina. The experiments reported briefly here provide additional evidence for this dual origin based on cell labeling. A more detailed consideration of these data will be presented in a separate publication.

DNA synthesis in pigmented cells of the iris epithelium which then failed to proliferate was observed not only in the absence of the neural retina, but also after lentectomy alone, in a few cells located some distance from the pupil. In a series of specimens fixed at successive intervals after injection, cells in this location remained heavily labeled while the label was being diluted by mitotic activity in the cells at the pupillary margin.

SUMMARY

When the lens was removed from the eye of adult *Notophthalmus viridescens* and the specific DNA precursor, thymidine-^{3}H, was injected intraperitoneally, the initiation of DNA synthesis was indicated by the appearance of labeled nuclei in the pigmented iris epithelium 4 days after lentectomy. As the cells near the mid-dorsal pupillary margin then undergo depigmentation to form a lens vesicle, DNA replication and mitotic cell division continue in this tissue. Cells which were labeled in the dorsal iris epithelium could be traced into all parts of the young, regenerating lens, including both the lens epithelium and the lens fibers. This radioautographic evidence, together with the histological evidence for continuity of the iris epithelium and the regenerating lens, provides support for the concept that dorsal iris cells transform into lens cells. As soon as the cells in the inner lamina of the lens vesicle begin to elongate to differentiate into primary lens fibers, DNA synthesis ceases in these cells but continues in the cells of the outer part of the lens vesicle and in the stalk connecting the lens vesicle with the rest of the iris epithelium. Cells in the latter areas continue to synthesize DNA and to divide, the daughter cells moving toward the primary lens fibers. In the equatorial zone, these cells begin to elongate, stop dividing and enclose the primary lens fibers with a shell of secondary lens fibers. As the lens grows in size, it detaches from the iris epithelium but the lens epithelium remains as a germinal epithelium providing a continuous supply of new secondary lens fibers.

Although DNA synthesis is initiated in the iris epithelium by 7 days after extirpation of both lens and neural retina, the other phases of lens regeneration are delayed until a layer of depigmented cells has been formed by the regenerating neural retina.

Unlabeled iris and lens vesicles were implanted into lentectomized eyes at different times after administration of thymidine-^{3}H to the host in order to test for the availability time of the isotope. Conspicuous labeling of the implant nuclei occurred up to two hours after injection and then dropped off so that no label was found from 3.5 hours to 14 days after injection. There was one exception where a light label appeared in a lens vesicle exposed to the host eye environment 2–3 days after isotope injection. It was concluded that the thymidine-^{3}H is available for only 2–3 hours in most instances.

I should like to thank Mrs. Evelyn Goldfein, Mrs. Joan Freed, and Mrs. Alice Barnett for their technical assistance in this research and Dr. Elizabeth Hay, Dr. Charles Leblond, and Mrs. S. Meryl Rose for suggestions concerning radioautographic techniques.

REFERENCES

Dumont, J. N., Yamada, T., and Cone, M. V. (1970). Alteration of nucleolar ultrastructure in iris epithelial cells during initiation of Wolffian lens regeneration. *J. Exp. Zool.* **174,** 187–204.

Eguchi, G. (1963). Electron microscopic studies on lens regeneration I. Mechanism of depigmentation of the iris. *Embryologia* **8,** 45–62.

Eguchi, G. (1964). Electron microscopic studies on lens regeneration II. Formation and growth of lens vesicle and differentiation of lens fibers. *Embryologia* **8,** 247–287.

Eisenberg, S., and Yamada, T. (1966). A study of DNA synthesis during the transformation of the iris into lens in the lentectomized newt. *J. Exp. Zool.* **162,** 353–367.

Eisenberg Zalik, S., and Yamada, T. (1967). The cell cycle during lens regeneration. *J. Exp. Zool.* **165,** 385–394.

Hasegawa, M. (1958). Restitution of the eye after removal of the retina and lens in the newt, *Triturus pyrrhogaster. Embryologia* **4,** 1–32.

Hasegawa, M. (1965). Restitution of the eye from the iris after removal of the retina and lens together with the eye coats in the newt, *Triturus pyrrhogaster. Embryologia* **8,** 362–386.

Hay, E. D., and Fischman, D. A. (1961). Origin of the blastema in regenerating limbs of the newt *Triturus viridescens.* An autoradiographic study using tritiated thymidine to follow cell proliferation and migration. *Develop. Biol.* **3,** 26–59.

Karasaki, S. (1964). An electron microscopic study of Wolffian lens regeneration in the adult newt. *J. Ultrastruct. Res.* **11,** 246–273.

Kopriwa, B. M., and Leblond, C. P. (1962). Improvements in the coating technique of radioautography. *J. Histochem. Cytochem.* **10,** 269–284.

Reese, D. H., Puccia, E., and Yamada, T. (1969). Activation of ribosomal RNA synthesis in initiation of Wolffian lens regeneration. *J. Exp. Zool.* **170,** 259–268.

Reyer, R. W. (1954). Regeneration of the lens in the amphibian eye. *Quart. Rev. Biol.* **29,** 1–46.

Reyer, R. W. (1962). Regeneration in the amphibian eye. *Symp. Soc. Study Develop. Growth* **20,** 211–265.

Reyer, R. W. (1966). DNA synthesis and cell movement during lens regeneration in adult *Triturus viridescens*. *Amer. Zool.* **6,** 329–330.

REYER, R. W. (1967). Lens regeneration following removal of the lens alone or after removal of the lens and neural retina. DNA synthesis and lens capsule differentiation. *Anat. Rec.* **157,** 308.

ROSE, F. C., and ROSE, S. M. (1965). The role of normal epidermis in recovery of regenerative ability in X rayed limbs of *Triturus*. *Growth* **29,** 361–393.

SATO, T. (1940). Vergleichende Studien über die Geschwindigkeit der Wolffschen Linsenregeneration bei *Triton taeniatus* und bei *Diemyctylus pyrrhogaster*. *Arch. Entwicklungsmech. Organismen* **140,** 570–613.

SCHEIB, D. (1965). Recherches récentes sur la régénération du cristallin chez les vertébrés. *Ergeb. Anat. Entwicklungsgesch.* **38,** 46–114.

STONE, L. S. (1950). The role of retinal pigment cells in regenerating neural retinae of adult salamander eyes. *J. Exp. Zool.* **113,** 9–32.

STONE, L. S. (1959). Regeneration of the retina, iris, and lens. *In* "Regeneration in Vertebrates" (C. S. Thornton, ed.), pp. 3–14. Univ. of Chicago Press, Chicago, Illinois.

STONE, L. S. (1960). Regeneration of the lens, iris, and neural retina in a vertebrate eye. *Yale J. Biol. Med.* **32,** 464–473.

STONE, L. S. (1965). The regeneration of the crystalline lens. *Invest. Ophthalmol.* **4,** 420–432.

STONE, L. S., and STEINITZ, H. (1953). The regeneration of lenses in eyes with intact and regenerating retina in adult *Triturus v. viridescens*. *J. Exp. Zool.* **124,** 435–468.

YAMADA, T. (1966). Control of tissue specificity: The pattern of cellular synthetic activities in tissue transformation. *Amer. Zool.* **6,** 21–31.

YAMADA, T. (1967a). Cellular synthetic activities in induction of tissue transformation. *Cell Differentiation Ciba Found. Symp.*, pp. 116–130.

YAMADA, T. (1967b). Cellular and subcellular events in Wolffian lens regeneration. *Curr. Top. Develop. Biol.* **2,** 247–283.

YAMADA, T., and ROESEL, M. E. (1968). Labeling of lens regenerate cells grafted into the newt optic chamber. A study of availability time of tritiated thymidine. *Exp. Cell Res.* **50,** 649–652.

YAMADA, T., and ROESEL, M. E. (1969). Activation of DNA replication in the iris epithelium by lens removal. *J. Exp. Zool.* **171,** 425–431.

DEVELOPMENT OF ^{3}H-THYMIDINE-LABELLED IRIS IN THE OPTIC CHAMBER OF LENTECTOMIZED NEWTS

SARA E. ZALIK and VI SCOTT

During lens regeneration in the newt *Triturus viridescens* DNA synthesis is activated in the pigment cells of the dorsal iris at about 4–7 days after removal of the original lens [1, 4, 7]. Long-term experiments in which regenerates were labelled with ^{3}H-thymidine and kept for different time intervals have shown that pigmented iris cells give rise to lenses in which all cell nuclei are labelled. Once depigmentation and cell elongation occur in the activated iris, injection of the isotope results in appearance of unlabelled lens fibers in subsequent regenerates. Although a number of studies suggest that ^{3}H-thymidine is available in the optic chamber for only about 3 h after injection [2, 3, 6], the possibility of delayed isotope incorporation in long-term labelling experiments is not excluded. In an attempt to overcome this, the activated dorsal irises were labelled and implanted into the optic chamber of unlabelled newts where subsequent regeneration occurred.

MATERIALS AND METHODS

Adult *Triturus viridescens* were used as donors and hosts and were kept at room temperature in individual plastic containers. Both eyes of the donors were lentectomized. Ten days later ^{3}H-thymidine (New England Nuclear Corp., spec. act. 6.7 Ci/mM, at a concentration of 10 μCi/g body wt, was injected intraperitoneally. Three hours after injection the dorsal iris of one lentectomized eye was removed washed in amphibian Ringer solution and implanted into the optic chamber of a host lentectomized immediately prior to receiving the labelled iris. The other eye of the donor newt was fixed in Bouin fixative and served as a control. Twenty explants were performed.

After 20 days the host animals were sacrificed and eyes containing the donor tissue were dissected and fixed. The tissues were embedded in paraplast using standard histological procedures, and sectioned at 5 μm. The slides were covered with Kodak NTB-3 liquid emulsion and exposed in the cold for 29 days. Following exposure they were developed with D-11, acid fixed and stained with hematoxylin-eosin.

For autoradiographic analysis, the number of grains per nucleus was recorded for the control irises. In the lens regenerates developed from the explanted irises, grain counts per nucleus were recorded for the lens epithelium, secondary and primary lens fibers, and pigmented and depigmented cells derived from the donor but not forming a lens. Only sections devoid of background or low in background were used. Nuclei having fewer than 5 grains were not considered labelled. Staging was performed according to Yamada [5].

RESULTS

Twelve of the twenty irises explanted developed into normal regenerates which ranged from stages X to XI. Control irises were at

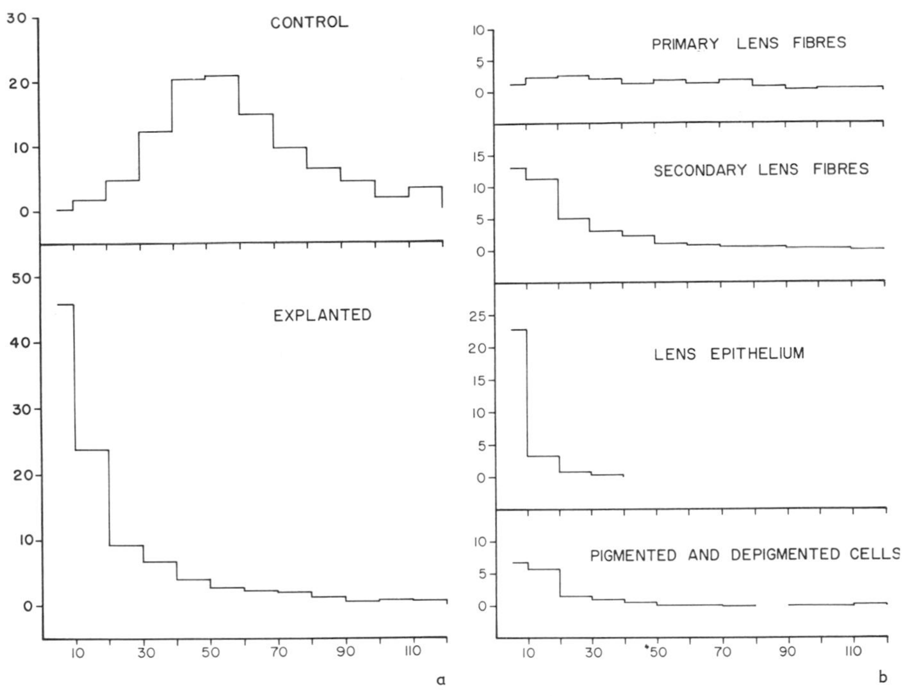

Fig. 1. Abscissa: number of grains/nucleus; *ordinate:* % labelled cells.
Frequency distribution of grain counts over individual nuclei in the control and explanted lens regenerates. A total of 940 cells were counted for the control irises and 1660 cells for the regenerates. (*a*) Frequency distribution for stages III to IV regenerates and explanted stages X-XI regenerates; (*b*) Frequency distribution of grain counts per nuclei in different regions of stage XI regenerates.

stages III to IV. The majority of the cells in the regenerates were labelled; however, in a few cases, more centrally located primary lens fibers were devoid of label.

The frequency distributions of grain counts per labelled nucleus in the controls and in the explanted lens regenerates are presented in fig. 1. It can be observed in fig. 1*a* that in the control irises about 70% of the nuclei possessed 30 to 70 grains, while in the implanted regenerates 45% of the population had 5–10 grains and 23% attained 11–20 grains per nucleus. The remaining 32% of the population had values of more than 20 grains per nucleus.

The contributions of the various regions to the total labelling pattern of the regenerates are presented in fig. 1*b*. Primary lens fibres represent a small proportion (16%) of the labelled cell population; however, these and the secondary lens fibres, mostly the inner zone of the latter, are the sole contributors of cells with 40 or more grains. Secondary lens fibres contributed 37% of the labelled cell population; of this 13% of the nuclei had 5–10 grains, 12% had 11–20 grains and 10% had 20–50 grains. The remaining 2% had values spreading from 50 to 100 or more grains per nucleus. The greatest proportion of nuclei in the lens epithelium had 5–10 grains, while a small number (3 %) attained 11 to 20 grains. Less than 1 % of the population had more than 20 grains. The majority of the cells with 5–10 grains per nucleus were present in the lens epithelium and the periphery of the secondary lens fibres. Some pigmented and depigmented cells from the donor irises were very closely attached to the explanted lens; however, they were not involved in its formation. Most of them possessed 5–20 grains per nucleus.

DISCUSSION

Although the experiments reported here do not rule out the possibility of an intracellular pool, they should have eliminated the possibility of delayed incorporation of the extracellular isotope present at low levels in the optic chamber. Our results demonstrate that cells of the stage XI regenerate are derived from the pigmented and partially depigmented cells of the dorsal iris at stage III. It would appear from our data that some cells giving rise to primary and inner secondary lens fibres were labelled at their terminal cell cycle. On the other hand, lens epithelial cells seem to be the actively dividing population, also derived from pigmented and partially depigmented dorsal iris cells. From the grain count data it appears that there is no delayed division of cells involved in the formation of the lens epithelium. These cells were evidently derived by division of the cells labelled at stage III. These experiments cannot resolve whether one daughter cell is involved in the lens fibre formation, while the other continues to divide as a proliferating stem cell, or both cells differentiate after the last S phase, and mitosis.

Labelled stage III regenerates were chosen as donor explants because the percentage of cells in DNA synthesis is highest at this stage [1]. A 3 h labelling time was employed since average grain count per labelled cell reaches a plateau after this time interval [2, 6].

The results of this study confirm our previous findings suggesting that the primary lens fibres, secondary fibres and lens epithelium are derived from the iris epithelium [1].

This work was supported by The Cancer Institute and the National Research Council of Canada.

REFERENCES

1. Eisenberg, S & Yamada, T, J exptl zool 162 (1966) 353.
2. Eisenberg-Zalik, S & Yamada, T, Ibid 165 (1967) 385.
3. Reedan, J R & Rothstein, H, J cell physiol 67 (1966) 307.
4. Reyer, R W, Coulombre, A J, Yamada, T & Papaconstantinou, J, Science 154 (1966) 1682.
5. Yamada, T, Current topics in developmental biology. (ed A A Moscona & A Monroy) vol. 2, p. 247. Academic Press, New York (1967).
6. Yamada, T & Roesel, M E, Exptl cell res 50 (1968) 649.
7. — J exptl zool 17 (1969) 425.

KEY-WORD TITLE INDEX

AUTHOR INDEX